Situated Complexity
Modeling Change
in Nonlinear Biological Systems
with a Focus on Africa

Dorothy Wallace

To order additional copies of this book, contact:
Xlibris Corporation
1-888-795-4274
www.Xlibris.com
Orders@Xlibris.com
72815

This text is dedicated
to the brave men and women
of Médecins Sans Frontières

Contents

CONTENTS

Preface

Modeling is our way of putting our understanding of science and medicine into a quantitative form. Only in this form can our knowledge be used to the make quantitative predictions about the future which form a basis for scientific or social action. The text you have before you is designed to guide a one semester course in modeling biological phenomena using systems of ordinary differential equations. It supposes that the student has seen a semester of calculus and knows the fundamental fact that the derivative of a function is its rate of change. It also assumes that the student has an interest in biology and medicine, as it draws on examples from both.

This text is not intended to be used in isolation, but as a primer leading to the research literature in biological modeling. The literature is growing rapidly, so none of the discussions of specific problems in the text should be taken as a final word on the subject. Even the references given to scholarly papers will be out of date by the time you read this. There will be new ones. The whole text is just a warmup exercise leading to the literature, which should be drawn into use as early in the course as possible.

Quotes from historical sources are now readily available via an internet search and so are not cited in the traditional way. Students are more likely to find old diaries and journals through an internet search than in the university library.

Very few problems discussed here can be approached effectively without the aid of a computer program capable of numerical integration. In my course at Dartmouth we use the Big Green Differential Equation Machine, available free on the web or a better downloadable version can be obtained for a small fee. Most of the graphical output seen in this text was done with Big Green. Many other software choices are possible but it is necessary to have some such program at hand to take full advantage of this text.

Modeling makes the most sense when models are constructed with respect to a particular problem. Biological, and especially medical, problems occur in a rich context of ecological, social, cultural and geographic factors. These factors may be irrelevant or crucial, depending on the exact nature of the problem under consideration. Figuring out which factors are relevant and how to take them into account is one of the most difficult aspects of modeling. This text focuses on issues taken from the context of Sub-Saharan Africa, in particular the region around Lake Victoria in East Africa.

The choice of context is highly intentional. Medical and ecological issues in this region are pressing and difficult to resolve, as well as being complex in cultural, biological and mathematical ways. American students know little about Africa and here is a chance to get to know one part of it. As they learn more about particular problems from the literature, their models can grow more complex to take important factors into account. The choice of problems is intended to create a certain moral discomfort with making easy decisions based on unquestioned assumptions built into overly simple models.

Acknowledgments

The author wishes to thank the National Science Foundation for its support of this project. In particular, Dr. Herb Levitan from the Foundation provided the initial impetus for the course, insisting repeatedly that something be done for biology students. Equally important was the continuing supportive atmosphere of Dartmouth College for trying out new courses and pedagogy, without which this text and the course it serves would never have happened.

The construction of this text has been from the user upward, in the sense that most of the topics included and the direction in which it developed were dictated largely by the research that undergraduates chose to do in the course that this text serves at Dartmouth College. Thus, the students themselves must be acknowledged as fundamental creators of this text. It is a luxury to teach from a text that contains topics I know are interesting to students, because they chose them.

The earliest version of the text owed much to Emily Cornell and Joanne Fixman for explanations of some of the simplest models, to Anne Margolis for research on water hyacinth, to James von Rittman for historical notes, and to Barack Abonyo for historical notes and a thorough description of his experiences with Kenyan fisheries. Heidi Williams deserves mention for a great chapter on the human development index that didn't make it into the final version. Erin Dauson found and organised references for most chapters. Mits Kobayashi and Jared Corduan contributed to the editing and finding of errors; Jared especially helped by doing the initial typesetting. Tom Shemanske fixed the errors none of the rest of us could fix. Maria Wallace designed the cover art. Thanks go to Professor

Rebecca Hartzler at Edmonds Community College, to Wendy Rockhill, Joshua Whorley and Bryan Johns at Seattle Central Community College, and to Professor Rachel Esselstein at California State University at Monterey Bay, for agreeing to try the text in their courses. Dr. Eluemuno Blyden was a particularly useful critic at the start of the project. Particular thanks go to Brian Reid for developing software (The Big Green Differential Equation Machine) that allowed my students to interact creatively with the problems in the text, and to Kim Rheinlander, whose evaluation of the course led to many refinements and for help with editing. Clyde Martin has been a tremendous resource for interesting problems bridging the two disciplines.

It is traditional to thank family and friends for support, and I do so here. The Dartmouth Mathematics department has staunchly supported the course that led to this text, and I am lucky to be able to include my colleagues as friends. I would certainly be remiss if I did not acknowledge explicitly the graciousness and support of my daughter Maria.

D. Wallace 2009

Magnificent Water

From the diary of John Speke:
"This magnificent sheet of water I have ventured to name Victoria, after our gracious Sovereign..."

The diaries of John Speke give away the tone of European enthusiasm upon discovering yet another confirmation of Arabic maps, geography and rumor. This particular rumor concerned the gigantic tropical lake that feeds the headwaters of the Nile river running north through Egypt. For centuries the continent of Africa had seemed impenetrable to Europeans, truly a "dark" continent hidden by a wilderness unknown to European maps and a collection of religions incomprehensible to northern Christians. In fact, the abbreviated history of European attempts to infiltrate various corners of Africa reveals many a wise and wary ruler forcing the white explorer back down the road whence he came. It is easy to imagine that the rumors of African wealth, wonders and dangers that white explorers heard from Arab sources had their reciprocal in equally terrifying stories of encounters with Europeans, delivered from those same Arabic sources to African ears, particularly stories of slavery. News of slavery reached into the interior of Africa, past European outposts, past Arab travel routes, right into the heart of the region. So, not surprisingly, the interior of Africa remained inaccessible for quite a long while in spite of repeated attempts of various European countries to colonize, conquer, or simply visit. Of course, what looks impenetrable and unwelcoming to the uninvited guest might be an open highway to the familiar resident. In fact, what earlier explorers were unable to accomplish through bravado, later ones could manage easily thanks to trade.

Trading patterns allow modern historians to piece together some of the details of medieval Africa. From A.D. 500 forward the region between Lake Victoria and the east coast of Africa saw advances in agriculture and ironworking, as well as exports of iron and ivory. The presence of imported items, such as the Chinese porcelain found in certain places, is also evidence of early trade with the coast. But the early coastal traders were careful to monopolize their trade with the interior of Africa by carefully concealing their suppliers of goods. Nonetheless, the remains of large cities in the region indicate an economy well sustained by trade, probably with coastal cities. The city of Engaruka, one of the last to grow during the medieval period, had a population estimated at between 30 and 40 thousand people. Midway between Lake Victoria inland and Mombasa on the coast, connected by established trade routes to the north and south, Engaruka was a flourishing city that waned long before John Speke's journey inland. Two main factors have been offered for its demise. The first was a series of invasions by nomadic peoples from the north. The second reason is largely economic. The trade which had sustained the growth of Engaruka vanished in the 1500's when the Portuguese and other European powers attempted to cut off the ocean trade to the mideast and Asia as they vied for control of trade along all of the African coast.

So, at the start of the 19th century the inland areas around Lake Victoria had very little contact with the coastal regions. Geographic barriers included tsetse infested forest, malaria ridden swamp, unnavigable rivers and wild animals. A lack of economic incentive to trade was also a large factor. While Arabian and Asian traders were interspersed along the length of the East African coast, they dealt primarily with resupplying ships in transit from Europe or Arabia. Trade between the coast and the interior remained insignificant. Coastal traders saw little within the interior worth trading for in light of the hazards of traveling.

Growing prices for ivory changed the economic calculus again around the beginning of the 19th century. As ivory prices rose with Asian and European demand, Indian elephants were hunted down and new sources for this "white gold" were needed. Arabian traders began to pay coastal Africans to make the trek inland to hunt elephants. As the elephants nearer the coast were hunted out these hunters went farther inland in search of their prey. This created a moving "ivory frontier" that would continue to migrate westward as the century progressed.

The Sultanate of Zanzibar, which controlled the whole of the East African coast, profited most from this trade. All trade was funneled through the island by placing monstrous tariffs on international trade conducted from the coast and other islands. Ivory was carried along trade routes running roughly east-west across Tanzania to the ports of Bagamayo or Dar es Salaam or, to a lesser extent Mombasa, greatly enriching these cities and the small, growing towns along the trade routes. Towns along the route were most often filled with an Asian or Arabian merchant elite and a larger African population for labor. As the century progressed, Christian missionaries, European civil servants and soldiers often joined the few non-Africans in these isolated towns.

The ivory trade originally operated through regional networks with the coastal merchants rarely entering the interior. Instead, the ivory was transported by an adventurous Arabian leading a small caravan of black Africans. The caravan would receive the ivory from an inland tribe hired to hunt elephants. As the caravan traveled, it would pay tribes whose lands it traversed for food and lodging.

Africans, seeing the profits to be made from hunting elephants, began to concentrate on this lucrative venture. Established farming communities began hunting for ivory and neglecting their fields. Other tribes were wholly enlisted as porters on the caravans. Midway into the 19th century, the ivory trade was so extensive as to employ more than 100,000 Africans traveling yearly on the central trade routes.

So, in 1858 John Speke was able to join a caravan and follow well travelled routes to the southern shores of the huge lake, where his first reaction was to give English names to the lake and the hill on which he stood.

"The caravan, after quitting Isamiro, began winding up a long but gradually inclined hill - which, as it bears no native name, I shall call Somerset - until it reached its summit, when the vast expanse of the pale-blue waters of the N'yanza burst suddenly upon my gaze. It

*was morning. The distant sea-line of the north horizon was defined
in the calm atmosphere between the north and west points of the
compass; but even this did not afford me any idea of the breadth
of the lake, as an archipelago of islands...each consisting of a single
hill, rising to a height of 200 or 300 feet above the water, intersected
the line of vision to the left; while on the right the western horn of
the Ukéréwé Island cut off any farther view of its distant waters
to the eastward of north. A sheet of water - an elbow of the sea,
however, at the base of the low range on which I stood - extended
far away to the eastward, to where, in the dim distance, a hummock-
like elevation of the mainland marked what I understood to be the
south and east angle of the lake. The important islands of Ukéréwé
and Mzita, distant about twenty or thirty miles, formed the visible
north shore of this firth...*

"*In consequence of the northern islands of the Bengal Archipelago
before mentioned obstructing the view, the western shore of the lake
could not be defined: a series of low hill-tops extended in this direction as far as the eye could reach; while below me, at no great
distance, was the debouchure of the creek, which enters the lake
from the south, and along the banks of which my last three days'
journey had led me. This view was one which, even in a well-
known and explored country, would have arrested the traveller by
its peaceful beauty. The islands, each swelling in a gentle slope to
a rounded summit, clothed with wood between the rugged angular
closely-cropping rocks of granite, seemed mirrored in the calm surface of the lake; on which I here and there detected a small black
speck, the tiny canoe of some Muanza fisherman. On the gently
shelving plain below me, blue smoke curled above the trees, which
here and there partially concealed villages and hamlets, their brown
thatched roofs contrasting with the emerald green of the beautiful
milk-bush, the coral branches of which cluster in such profusion
round the cottages, and form alleys and hedgerows about the villages
as ornamental as any garden shrub in England. But the pleasure of
the mere view vanished in the presence of those more intense and
exciting emotions which are called up by the consideration of the
commercial and geographical importance of the prospect before me.*

"*I no longer felt any doubt that the lake at my feet gave birth* to that interesting river [the Nile], the source of which has been the subject of so much speculation and the object of so many explorers. The Arabs' tale was proved to the letter. This is a far more extensive lake than the Tanganyika; 'so broad you could not see across it, and so long that nobody knew its length.' I had now the pleasure of perceiving that a map I had constructed on Arab testimony, and sent home to the Royal Geographical Society before leaving Unyanyembé, was so substantially correct that in its general outlines I had nothing whatever to alter. Further, as I drew that map after proving their first statements about the Tanganyika, which were made before my going there, I have every reason to feel confident of their veracity relative to their travels north through Karagué, and to Kibuga in Uganda.

"...'If you have come only to see a large bit of water, you had better go northwards and see the Ukéréwé; for it is much greater in every respect than the Tanganyika;' and so far as I can ascertain, it is. Muanza, our journey's end, now lay at our feet. It is an open, well-cultivated plain on the southern end, and lies almost flush with the lake; a happy, secluded-looking corner, containing every natural facility to make life pleasant...

"...I further draw my conclusions from the fact, that all the hills on the country are much about the same height - two or three hundred feet above the basial surface of the land; and I could only see the top of the hill like a hazy brown spot, contrasted in relief against the clear blue sky. Indeed, had my attention not been drawn to it, I should probably have overlooked it, and have thought there was only a sea horizon before me. On facing to the W.N.W., I could only see a sea horizon; and on inquiring how far back the land lay, was assured that, beyond the island of Ukéréwé, there was an equal expanse of it east and west, and that it would be more than double the distance of the little hill before alluded to, or from eighty to one hundred miles in breadth.

"On my inquiring about the lake's length, the man faced to the north, and began nodding his head to it: at the same time he kept throwing forward his right hand, and, making repeated snaps of his fingers, endeavoured to indicate something immeasurable; and added, that nobody knew, but he thought it probably extended to the *end of the world"*.

At 68,460 square kilometers, Lake Victoria is approximately the size of Ireland and three times the size of New Hampshire. Situated at 1136 meters above sea level, it is the world's largest tropical lake and the second largest freshwater lake. As Speke discovered, Lake Victoria is the source of the White Nile. Out of proportion with the lake's large size is its depth, which never reaches more than 84 meters. Its mean depth is 40 meters, which is extremely shallow for a lake of this size. Putting its depth into perspective a little more, 40 meters is not even the length of an Olympic swimming pool. Of course, none of these facts were available to Speke when he first viewed the lake. Rather, what struck him were the attractive possibilities for colonizing the region, the potential for economic development, and the sheer attractiveness of the pale blue waters of the lake.

That blueness would eventually disappear. Although we usually associate pristine blue waters with healthy lakes and streams, the truth is that blue water is not very fertile. Algae require various nutrients to grow, and without them cannot reproduce very quickly. All of the life in the lake depends on the productivity of its algae. As with many lakes, the ecosystem of Lake Victoria was based on complex interactions among species in many trophic levels (see Figure 1). At the bottom of the food web were producers (organisms which convert energy into a form that can be used by other organisms) such as algae and diatoms.

There were several fish and animal species in and around the lake which directly or indirectly depended on the algae population. Even the bones of Pleistocene Nile perch which people regard as a strictly newly introduced species have been found on Rusinga island. The major fish species found here before the introduction of Nile perch include *omena*(sardine, *Rastreoneobola argentia*), *kamongo* (Lungfish, *protopterus sp.*), *mumi* (catfish, *clarius mossambicus*), *ningu* (catfish, *cyprinus carpio*), *odhadho* (catfish, *Labeo*), *okoko* (catfish, *bagrus docmac*),*fulu* (*Haplochromis sp.*), *ngenge* (Tilapia, Tilapine *sp.*), eel and others. The number of *mbuta* (Nile perch, *lates niloticus*) had dwindled and dissapeared for reasons which have not yet been described by scientists. Other animals such as otter, a few

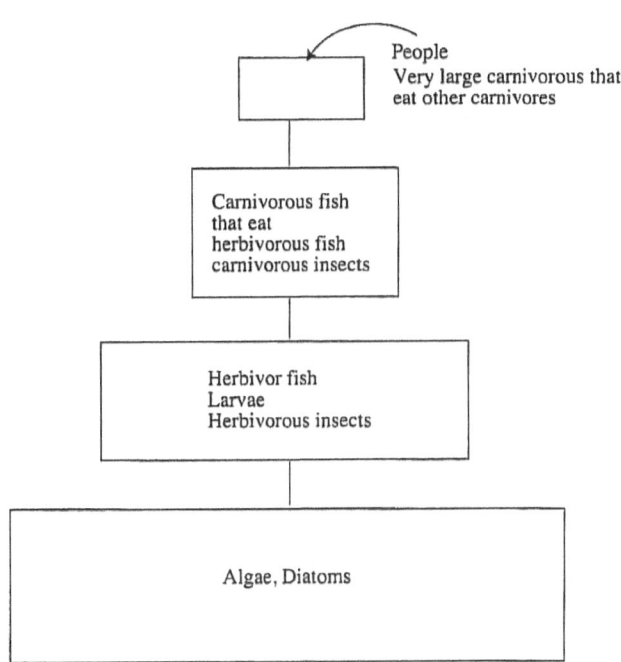

Figure 1: Simple trophic levels.

crocodile, hippopotomus, and fish eating birds also constituted the ecology of this lake. The food chain is complex, with smaller fish species such as sardines and haplochromis feeding on phytoplanktons (blue green algae, diatoms etc), zooplanktons (copepods) and dead decomposing aniamal and plant material (detritus). The pattern of feeding by tilapine species is rather complex since in addition to zooplankton and phytoplanktons some fish fingerlings, insects larvae, molluscs and leaches have been found in their gut. Most of the catfish species are piscivorous but may also feed on prawns, dragonfly and other insect species. The otters,crocodiles and other birds feed on a variety of fish species. The relation between the hippotomus and the fish is strictly symbiotic. Hippos only feed on live folliage and their excrements may constitute the detrital material upon which several fish species depend. Man is at the top of the food chain feeding on both plant and animal material including the crocodile and the hippotomus. Haplochromis constituted the highest population of the lake followed by tilapia.

What the lake had in abundance was warmth and sunlight. What it had in very small amounts were the nutrients needed to support a larger algae base for the food chain.

In the early 1800s, partly as a result of the need for raw materials and markets for manufactured goods created by the Industrial Revolution, Europe began to increase its trade with the East African coast. Seyyid Said, ruler of the part of this coast that included Oman, Mombasa, and Zanzibar, made trade treaties with Great Britain, France, and the United States in the 1830s and 1840s which heavily favored Europe and subordinated East Africa. Missionaries, especially during the period from the 1870s through the 1890s, increased European influence in East Africa. A strong British anti-slavery movement which saw slavery as inhumane, unChristian, and economically unsound, further increased European influence over East Africa. Said had no choice but to accept treaties with Great Britain in 1822 and 1845 which restricted the East African slave trade. Said's successor, Sultan Barghash, was then forced to accept the abolition of the slave trade in 1873.

In the mid-1880s Germany made part of East Africa a protectorate and in 1886 Britain and Germany signed an agreement that partitioned the section of East Africa from the east coast to the shore of Lake Victoria. The northern part of this area was under Great Britain's sphere of influence while Germany gained rights to the southern part. These areas more or less correspond to Kenya in the north and Tanzania in the south. Since Britain was worried that another European power could threaten her control of Egypt by damming the Nile, in 1890 another agreement between Germany and Britain divided Lake Victoria as well. The British again took the land in the north (modern Uganda) and the Germans that in the south. Chartered companies were founded by Europeans to take over and govern the new territories in East Africa but they were largely unsuccessful. In 1894 and 1895 Great Britain was forced to create protectorates, one over Uganda and another over the rest of its East African territories (called the East Africa Protectorate). Germany had already taken similar actions in 1890 and 1891.

European powers tried to get the colonies to support themselves. The economic systems that had worked for centuries in that region were never built to sustain a colonial lifestyle imported from England or Germany. At least one historian has privately noted the resemblance of Africa's colonists to the top predator in the food chain of Lake Victoria. At the risk of straining the analogy, a system that can support many farmers, weavers, potters and metalworkers at a modest level strains to support more than a few wealthy landowners, a small platoon of soldiers and a handful of missionaries.

In the Lake region the European came first as a religious man interested only in improving the lives of the local community by providing education and other enlightenment. There was no slave trade. However the soft spoken religious whites were soon replaced by the colonial white government officials who thought African land was up for grabs. They secured certain parts of the Lake region and other parts of the highlands far and beyond this region where they introduced intensive farming. Of course, Africans were the laborers at meagre pay or none except for food and housing. Crops such as wheat, sugarcane, cotton, coffee,and tea soon sprang up in massive borderless farms. Nitrates, sulphates and other chemicals that have just recently been found to be hazardous to the environment were used in these farms and all of them ended up in

Lake Victoria through the rivers winding through the region. Industries grew to process the raw material. Many towns like Kisumu in Kenya, Mwanza in Tanzania, Jinja in Uganda and others sprang up around the lake. These activities, coupled with the fisheries themselves, led to immigration of different ethnic groups. For example, the Kikuyus, Kalenjins, Kambas, Durumas, Embus, Gusis, Indians, Pokomos,Turkanas, basically every one of the thirty four tribes of Kenya are represented, not to forget other foreigners who have come to invest in the region.

Over the course of years, all of this growth and activity had the overwhelming result of pouring large amounts of nutrients into the clear blue waters of Lake Victoria. The algae population suddenly found itself in possession of a nearly infinite supply of foodstuffs, and it began to grow accordingly. As a result of European colonization of the region surrounding Lake Victoria, the number of humans living near the lake grew tremendously. Industrial effluent and agricultural chemicals (which washed from farms into tributary rivers) introduced large quantities of nutrients into Lake Victoria. Untreated sewage not only contributes additional nutrients, it also poses a major threat for transmission of diseases such as typhoid, cholera, and diarrhea. A study made around 1995 found that 2 million liters of untreated sewage and industrial effluent from Tanzania alone was flowing into Lake Victoria daily.

Now, at the start of the 21st century, we can quantify and study phenomena like these using the calculus of differential equations. A differential equation is simply a relationship among some quantities and their rates of change. In this text we will look at "ordinary" differential equations, which are those whose rate of change is always with respect to one single variable, in this case time. We will look at quantities that naturally change over time, such as populations of organisms, prevalence of disease, and so forth. The calculus of differential equations will allow us to state our understanding of how systems work in a mathematical form. We can use a computer to calculate numerical solutions to our equations and display them in various graphical forms, so that we can see what effects follow from the relationships we believe govern the system.

Chapter 1

Blue Lake, Green Lake

When ecologists study an ecosystem as complex as the one in Lake Victoria, they often try to write mathematical models to describe some of the interactions that they see. They hope to use these models to predict future changes in this or a similar ecosystem or, alternatively, to hypothesize about the conditions of the ecosystem in the past. After studying an ecosystem for some time, an ecologist may be able to predict which factors are most important to the specific system. At least at first, one may make certain assumptions about the system and its interactions in an attempt to make it more manageable to model. For instance, one may assume that the system is closed (there is no gain or loss of resource or biomass). In some systems this is a reasonable, albeit not completely accurate assumption at least for a small amount of time. Even if a system is not closed, there may be a negligible net flow of resource. The amount gained may approximately equal the amount lost.

An ecologist may try to represent pieces of the ecosystem separately and then later try to incorporate some of these many interactions into a more general model. In doing so, one is able to model specifically how a few organisms interact and gain some understanding of how different elements of the system work before the model gets too complex to be usefully analyzed. The ecologist may try to simplify some of the models by making assumptions that certain conditions are true (at least during a reasonable interval of some other factor such as time), or try looking at a group of organisms as one organism (especially if they all belong to the same trophic level) to see what the overall system will tend to do.

If we want to model the observed algae bloom in terms of growth of the population of algae in Lake Victoria, we might start by making some simplifying assumptions:

1) Resources (such as essential nutrients) are unlimited. This is a reasonably valid assumption at least for a small amount of time when the amount of available resource is large relative to the population of algae.

2) There is no migration of algae into or out of Lake Victoria.

3) The rates of reproduction and death (natural and due to predation) are constant.

If N represents the population of algae then the change in the population of algae (ΔN) during a certain period of time (Δt) is then simply the number of algae added to the population (B) by reproduction minus the amount that are taken out of the population (D) by death. Symbolically:

$$\frac{\Delta N}{\Delta t} = B - D$$

B represents the total number of algae produced over a period of time. Because the more algae there are, the more algae can be grown, B is proportional to the population (N) of algae. Thus B is found by multiplying b, the average number produced per algae over this time period, by the number of algae (N):

$$B = bN$$

Similarly,

$$D = dN$$

where d is the fraction of the algae population that dies during this time period. Then by substitution,

$$\frac{\Delta N}{\Delta t} = bN - dN$$

The change in the population of algae over a given period of time, ($\frac{\Delta N}{\Delta t}$), according to this model, is proportional to the difference between the birth rate (b) and death rate (d). Mathematically we can represent this as:

$$\frac{\Delta N}{\Delta t} = (b - d)N$$

The algae population has increased over the last 50 years in Lake Victoria and therefore we can deduce that the birth rate of algae is greater than its death rate. So, ($b - d$) is positive and can be replaced by a positive constant, r, which will represent the overall rate of population increase per algae in the lake:

$$\frac{\Delta N}{\Delta t} = rN$$

If we want to determine how the population of algae is changing at a particular time (determine its instantaneous rate of change) we would take the limit of $\frac{\Delta N}{\Delta t}$ as the period of time gets infinitesimally small to obtain the derivative of N with respect to time, resulting in the differential equation:

$$\frac{dN}{dt} = rN$$

In mathematical terms this equation states that the rate of change of the population at a given time is directly proportional to the population at that time. In population studies, a differential equation usually describes relationships between a function and its derivatives, and gives us a way of mathematically representing what rules we believe govern how a function such as population changes over time. Sometimes the equations enable us to see trends in a function even if we are unable to determine the function itself.

In the situation we are studying, $\frac{dN}{dt}$ represents the rate of change of the population of algae (N). If $\frac{dN}{dt}$ is positive, the rate of change of the population is positive and therefore the algae population is growing. Similarly, if $\frac{dN}{dt}$ is negative, the size of the population of algae is decreasing. The magnitude of $\frac{dN}{dt}$ tells us how quickly the size of the population is changing.

If we can figure out how the population, N, changes with time, then we can determine the size of the algae population at any time,

t. We want to find the function N in terms of t which means that we must solve the differential equation

$$\frac{dN}{dt} = rN$$

There are several strategies we can use for approaching a model of this sort. One is to attempt to find what scientists call an "analytic" solution to the problem. This means finding all the functions, $N(t)$ which solve the differential equation describing our model. Such a solution is very satisfying and allows one to analyze both the implications and the reliability of the model quite efficiently. Most biological models are too complicated to solve analytically, but the model we are using for algae growth is so simple that an analytic solution is very easy to calculate.

Using separation of variables we rearrange the terms of the equation:

$$\frac{dN}{N} = rdt$$

We then integrate:

$$\int \frac{dN}{N} = \int rdt$$

to obtain

$$ln|N| = rt + c$$

Since N is a population size and is therefore always non-negative, we may remove the absolute value:

$$lnN = rt + c$$

Next, we solve for N:

$$N = e^{rt+c}$$

The size of the algae population varies depending on when it is observed.

N is a function of time so we can represent N as $N(t)$. Solving for $N(t)$:

$$N(t) = e^{rt+c}$$

$$N(t) = e^{rt}e^{c}$$

At the initial time $t = t_0 = 0$, the population of algae is some value, N_0. So,

$$N(0) = N_0 = e^0 e^c = e^c$$

$$N_0 = e^c$$

Finally,

$$N(t) = e^c e^{rt} = N_0 e^{rt}$$

An equation in this form is commonly known as the Malthusian exponential growth model, named after the man who first proposed it as a model for human population growth at the end of the 18th century. If there are sufficient nutrients in Lake Victoria (because of pollution from industry and agriculture) we would predict that this model would give us a fairly good approximation of what is really happening to the algae. If surplus nutrients continue to be added to the lake and if this model is an accurate model of the algae in Lake Victoria, we should expect a continued exponential increase in the number of algae over time.

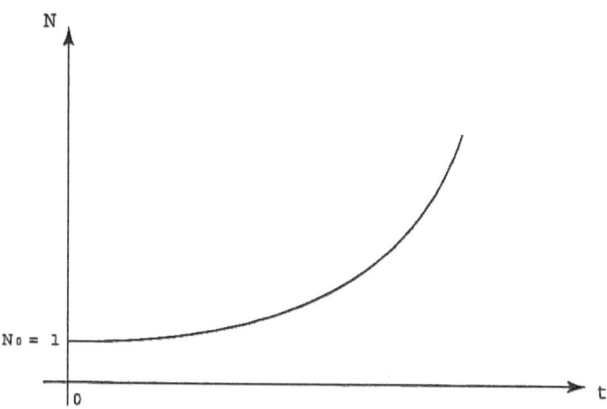

Figure 1.1: $N(t) = N_0 e^{.054t}$

According to some sources, between 1950 and 1980 the algal biomass production had increased somewhere from five- to ten-fold in the open surface waters. Using our model, if there is a five-fold increase in algae over 30 years, we represent the initial amount of algae as x and the five-fold increase as $5x$:

$$5x = xe^{30r}$$

$$5 = e^{30r}$$

which gives

$$r = .054$$

So,

$$N(t) = N_0 e^{.054t}$$

A graph of this equation if we choose $N_0 = 1$ is shown in Figure 1.1.

If the algae population does actually increase by nearly 50 times its original size by the year 2000 there may be many problems in store for Lake Victoria. Algae blooms not only effect the transparency of the lake water, but they have already greatly increased the mass of decaying matter and resulted in an increase of decomposer bacteria. The use of oxygen in the decomposition process contributes to deoxygenation of the lake, especially in the deeper waters. Less oxygen makes it more difficult for many fish species to survive and asphyxia has become a problem in some of the deeper parts of the lake. Upwelling of deoxygenated water has caused some fish kills. Due to the deoxygenation of parts of the lake, fish are being forced to crowd into the more oxygenated shallow regions, in turn draining the oxygen supply there. In addition, fishing effort has traditionally been concentrated more in the shallow waters of the lake so when fish are forced into shallower waters, they also face increased risk of being caught.

Since the growth of the algae population in Lake Victoria seems to have such negative effects, we may ask whether this growth will ever cease. An arrest in growth means that the size of the population remains constant. According to this model, are there any situations in which the size of the algae population remains constant? In other words, is there any number of algae for which the population is in equilibrium?

Equilibrium is a condition in which the rate of increase equals the rate of decrease. This means that the overall rate of change is zero. Symbolically,

$$\frac{dN}{dt} = 0$$

Where is this true for our model? To find out, we set

$$\frac{dN}{dt} = 0 = rN$$

Since r is always positive in our model,

$$N = 0$$

The only situation for which the size of the population of algae is not changing is when there are no algae ($N = 0$). Since in this model r is always positive, for any number of algae greater than zero the population will increase exponentially. The algae population will then continue to increase indefinitely (Since r is positive, $\frac{dN}{dt}$ will always be positive for this model). Thus $N = 0$ only occurs when there is no algae population to start with ($N_0 = 0$). The population must then remain at zero because it is not possible for algae to be created when there are not any to reproduce in the first place.

Analysis of equilibrium points sheds some light on the weaknesses of the Malthusian model. In reality, populations do not grow without bound. Mathematically, this means that a more accurate model might afford a variety of equilibrium points for the population, depending on the starting circumstances. A better model might afford other sorts of stable behavior too, such as oscillations in population size, which are also observed in nature. Nonetheless, the model has its uses, one of which is predicting future human populations, which are assumed to be in possession of unlimited resources. Forecasters are particularly fond of making predictions about African populations in particular. These predictions are generally based on the same model we just used for algae growth, and result in international policies concerning such things as birth control and deployment of associated medical personnel. The Malthusian model of exponential growth controls a lot of international opinion about the state of Africa.

For additional sources on first order algae growth, with a deeply scientific explanation, see Geiderv & Platt (1986). The assumption that nutrients are unlimited is not always valid. Instead of algae blooms assuming unlimited essential nutrients, as in this chapter, algae are also affected by other means such as amount of light present. Light is not the defining factor in Lake Victoria but Lisi & Totaro (2007) demonstrate how light can affect algae growth in other lakes, and models this process. Lake Victoria is not unique in its relations to algae blooms. For a look at another large lake, closer to home, see Hutchinson (1961), which models algae growth in Lake Erie, and explores algae biodiversity in relation to nutrients.

For your consideration

Question 1:

From the model of algae growth and the rates given above, what can we predict about how large the algae population of Lake Victoria will be over a 50 year period? A range is possible, varying from the low rate of a 5-fold increase in 30 years to a high 10-fold increase in 30 years. Of course scientists have studied what algae really do in tanks and in the open. Go to the literature and see what growth rates have been observed for various algae species in various environments. Do these match the ones given in the text? How large is the error in estimating the growth constant? The other critical constant in this model is the starting value of algae. How big an error would that estimate be likely to have? Now you can ask an important modeling question: Which error will create a bigger error in the predictive capacity of the model? To answer this question you need a way to make a comparison, both between two kinds of error and also two kinds of data collection. Furthermore you need a way to display your conclusion in a convincing graphical manner.

Question 2:

Go online and find population data for Kenya, Tanzania and Uganda. According to the population tables for the three countries, what is the range of possibilities for population growth rates? In which of the three countries is the population growing the fastest? According to the growth rates you calculate, what will the population of Africa be 20 years from now? 50 years from now? According to the growth rates you calculate, how many years ago was the population of each of these three countries less than 100 individuals? Of course, your answer will be a range of years reflecting the range of growth rates you calculated for each country. Now, compare this answer with what the literature says about the duration of human existence in Africa. If the model is inconsistent with the literature, then what is it telling you about instead?

Question 3:

Many people believe the population growth rate in African countries is dangerously high. They use a calculation similar to the one you used in question 2 to justify this argument. They are worried that Africa's population will outstrip its resources (especially food) in the near future, resulting in widespread starvation and disease. Other people believe that the rate of population growth one calculates from twentieth century data (such as you used) does not reflect longer term rates of growth. They use a calculation similar to the one you used in exercise 4 to argue that the rates calculated from recent data lead to an absurdly recent date of human origen in the region. One such argument is presented after the population tables at the end of this chapter. What do you believe? How do you reconcile these two different interpretations of the Malthusian model?

The following letter, no longer found on the internet, used the Malthusian model backwards to argue against claims made about population growth in Africa. Analyze the argument in this letter. Is it correct mathematically? What are its implications? An interesting research question would be to try to find data for the rate at which new languages and dialects were created, and then use that rate to estimate when there was only one language in a region.

A letter about African populations, condensed and rephrased

In just one relatively small country – Senegal – the Summer Institute of Linguistics, Inc. lists 38 distinct languages indigenous to the nation. Senegal is just a small part of Africa – the land mass of the continent as a whole is equal to 154 times the space in Senegal, and virtually all of it boasts similar ethnic diversity. The issue of language makes an interesting point.

Senegal's population is estimated at 9 million today. Supposedly, Senegal is one of those country's whose yearly population growth rate is above three percent. At a three percent yearly rate of population increase, Senegal would have had about 20,000 people at the start of the 19th century to arrive at 9 million today.

Now those 38 languages mentioned above are ancient. If anything, there are far fewer languages today than there were in centuries past. Any anthropologist will tell you that unique local languages are dying out – not being invented. So are we to believe that 20,000 people spoke 38 different languages in the year 1800?

The point is that the diversity of ethnic groups among a Senegalese population of nine million today suggests that these ethnic groups individually are very small. This is not consistent with a history of rapid population growth.

Chapter 2

Know before you go

From the Journal of Mungo Park:

"August 12th.–Rained all the morning. About eleven o'clock, the sky being clear, loaded the asses. None of the Europeans being able to lift a load, Isaaco made the Negroes load the whole. Saddled Mr. Anderson's horse; and having put a sick soldier on mine, took Mr. Anderson's horse by the bridle, that he might have no trouble but sitting upright on the saddle. We had not gone far before I found one of the asses with a load of gunpowder, the driver (Dickinson) being unable to proceed (I never heard of him afterwards); and shortly after the sick man dismounted from my horse, and laid down by a small pool of water, refusing to rise. Drove the ass and horse on before me. Passed a number of sick. At half past twelve o'clock Mr. Anderson declared he could ride no farther. Took him down and laid him in the shade of a bush, and sat down beside him. At half past two o'clock he made another attempt to proceed; but had not rode above an hundred yards before I had to take him down from the saddle, and lay him again in the shade. I now gave up all thoughts of being able to carry him forwards till the cool of the evening; and having turned the horses and ass to feed, I sat down to watch the pulsations of my dying friend. At four o'clock four of the sick came up; three of them agreed to take charge of the ass with the gunpowder; and I put a fourth, who had a sore leg, on my horse, telling him if he saw Mr. Scott on the road to give him the horse."

Early European explorers may have relied on Arabic maps and advice in their travels, but they didn't have the advantages of information that we now enjoy. Travel guides, tourist boards, and government web sites all offer a wealth of both friendly advice and warnings to those contemplating a visit to Africa. One consistent theme of these warnings is the possibility of getting sick. Visitors to a new place always run the risk of getting sick, just because the variants of local colds, flus, and dysentery are different from those at home and so the visitor hasn't yet acquired immunity to these strains. In the U.S. we have a name for this phenomenon: "turista", referring to the intestinal disorders U.S. residents often get when visiting Mexico. Of course, Mexican residents will have the same problem if they visit the U.S. But the word "turista" is woefully inadequate to describe the variety and intensity of tropical diseases available to visitors in Africa.

Those from prosperous northern countries live in communities where childhood vaccinations, availability of antibiotics, and a general lack of communicable diseases make us more aware of those ailments associated with genetics, lifestyle, and advancing age. It is hard for those of us from North America and Europe to imagine the incidence and variety of good old-fashioned infections that occur in tropical regions.

The World Health Organization lists an alarming number of diseases to think about when visiting Africa. In addition to familiar diseases such as malaria, HIV, and cholera, there are many less familiar ailments such as sleeping sickness (trypanosomiasis), dengue fever, ebola, Marburg haemorrhagic fever, schistosomiasis, Rift Valley fever, and yaws. Some diseases are transmitted by insects, some by human contact. Vaccines or other preventative measures exist for some of them, but not for most. Treatments exist for some but not all. Many are hard to diagnose in initial stages but deadly in later stages. The World Health Organization struggles with the problem of managing the epidemiology of a large number of diseases over a huge geographic region with many people in it, few of which have resources to pay the cost of a vaccine or a cure (if either exists).

From a letter to Margaret Sewell from David Livingstone,
Sept 9, 1850:

"When we got to the Ngami we found the fever raging. A fever
I suppose not very unlike that which cut up the Niger expedition.
It seems destined by Providence to keep the Intertropical Africa for
the black races alone."

It is probable that disease played a major role in disrupting early
attempts to colonize or even trade with African peoples. In 1805
Mungo Park took 44 Europeans on an expedition to discover the
source of the Niger River. All but 4 died. In 1816 James Tuckey
tried to explore the Congo and lost 37 % of his company. In 1832
M'Gregor Laird took 48 men on two ships to explore the Niger delta.
Nine survived. In 1841 Trotter took ships up the Niger carrying a
mixed crew including 145 Caucasian Europeans and 133 Africans
from Sierra Leone. 130 of the Caucasians got sick and 50 died,
but none of the African crew got sick. (from Livingstone's letter
to McWilliam, 1860) In every one of these examples the explorers
reported how disease took the lives of their company. In these
instances the diseases of the area actually protected its inhabitants
from (largely) unwanted trade or colonization.

It is likely that modern human beings existed first in East Africa.
The famous Leakey fossils represent one kind of archeological evidence. Fossil evidence is now backed by DNA studies that
demonstrate a higher level of genetic diversity in current African
populations than elsewhere, which is generally considered evidence
of a longer evolutionary period for the species. It is reasonable to
expect that the cradle of humanity would also be the cradle of its
most tenacious and adaptable diseases. Fortunately for those of us
who live in colder areas, many of these diseases are highly special-
ized to make use of insect vectors or animal reservoirs that are only
present in warmer areas. Projections of global warming, however,
indicate that the region affected by all of these diseases is likely to
grow.

All of these diseases present classes of interesting problems to the
modeler. There are basically three kinds of questions to consider:
epidemiology, physiology and pharmacology. Of course these are

related. Understanding the physiology of a disease leads to better understanding of how to prevent transmission of it, which informs epidemiology models. Treatment of disease (including pharmacology) also depends on understanding its physiology. But as models these three kinds of investigations are usually handled separately.

To model the physiology of an infection we have to understand what kind of parasite is involved, what its life cycle inside the body looks like, where it lives and what it damages, where it can hide. Some diseases, such as trypanosomiasis, have separate stages that differ in what organs of the body are affected, how severe the affliction is during that stage, the death rate, and whether a particular drug or other intervention will work. Some diseases have characteristic symptoms that should be predicted by an accurate model. A model that incorporates the correct life cycle of a parasite during an infection should be able to predict symptoms characteristic of that disease, such as the periodic temperature spikes characteristic of malaria infection. A model that correctly predicts the progress of the disease confirms our understanding of its physiology, while a model that expresses well-understood physiology mathematically is useful for suggesting interventions and epidemiology strategies.

From a letter to James McWilliam from David Livingstone, 1860, on fever in the Zambesi:

"The prescription employed is –Resin of jalap, and calomel, of each eight grains; quinine and rhubarb, of each four grains; mix well together, and when required make into pills with spirit of cardamoms: dose from then to twenty grains. * *The violent headache, pains in the back, etc., are all relieved in from four to six hours; and with the operation of the medicine there is an enormous discharge of black bile, – the patient frequently calls it blood. If the operation is delayed, a dessert-spoonful of salts promotes the action. Quinine is then given till the ears ring, etc. . . . . Those who may try the remedy will do well to remember that the above doses are for strong adults."*

Pharmacology concerns the invention and implementation of drug therapy for a disease. One branch of pharmacology looks at how drugs pass through the body: how they are absorbed, metabolized and eliminated. These processes are all described by models, which give a basis for assigning dosages and frequencies for any given medication. Some of the tropical infections require strong medications that have serious side effects. Keeping the concentration of drug in the bloodstream at an effective but non-toxic level is an important mathematical problem.

Epidemiology is the study of how a disease propagates throughout a population. The entire field makes extensive use of mathematical and statistical models to predict in advance whether a particular intervention will be effective. "Effective" in this case could mean different things, such as reduction of hospital caseload or the actual stopping of an epidemic. Models must take into account the methods of transmission of the disease, including to and from any animal vectors or reservoirs. The tropical diseases feature a lot of insect vectors, including flies, mosquitoes, and snails. Often the life cycles of these insects must be included in a model in order to answer basic epidemiology questions. Sometimes the most effective intervention is made on the insect population rather than the human population. Sometimes the cost of an intervention is built into the model as an additional point of comparison, and a particularly relevant one for poorer areas. Only through modeling can various proposed strategies be compared with one another before implementing them.

From the Journal of Mungo Park

"October 28th.–At a quarter past five o'clock in the morning my dear friend Mr. Alexander Anderson died after a sickness of four months. I feel much inclined to speak of his merits; but as his worth was known only to a few friends, I will rather cherish his memory in silence, and imitate his cool and steady conduct, than weary my friends with a panegyric in which they cannot be supposed to join. I shall only observe that no event which took place during the journey, ever threw the smallest gloom over my mind, till I laid Mr. Anderson in the grave. I then felt myself, as if left a second time lonely and friendless amidst the wilds of Africa."

Would Mungo Park have undertaken his expedition if he had known that 40 out of his 44 men would die? How long will the famous AIDS "cocktail" stave off the inevitable decline of T-cells in a patient? If the World Health Organization succeeds in putting a mosquito net over every African child at night, what will this do to reduce the death rate of malaria? Modeling has value at every point in our consideration of a disease, from our initial understanding of its physiology to high-level policy decisions about its control.

Chapter 3

Tsetse

For almost 20 years, starting in 1986, the citizens of Uganda suffered from ongoing civil war. Numerous factions battled in various regions of the country, with 400,000 people left homeless. Near the start of the unrest Doctors Without Borders established a project in northern Uganda. The mission of Doctors Without Borders is to provide aid to "people whose survival is threatened by violence, neglect, or catastrophe, primarily due to armed conflict, epidemics, malnutrition, exclusion from health care, or natural disasters." Uganda citizens clearly experienced more than one of these conditions, many of which are linked explicitly to increased transmission rates for infectious diseases.

In particular, Uganda was ripe for a resurgence of sleeping sickness. According to the World Health Organization, "The rural populations living in regions where transmission occurs and which depend on agriculture, fishing, animal husbandry or hunting are the most exposed to the bite of the tsetse fly and therefore to the disease. Sleeping sickness generally occurs in remote rural areas where health systems are weak or non-existent. The disease spreads in poor settings. Displacement of populations, war and poverty are important factors leading to increased transmission." Doctors Without Borders began treating patients with sleeping sickness through its northern Uganda project. Because of the prevalence of the disease, the organization eventually became responsible for supply and distribution of all sleeping sickness drugs in use worldwide today.

In 2007 they treated one case in 5 worldwide. Since the first cases in Uganda, the organization has screened a total of more than 2.4 million people for this disease and has treated over 43,000. As a point of comparison, in 2009 the World Health Organization estimated that 50,000-70,000 people were known to have the disease worldwide, although they estimated that 300,000 to 500,000 cases remained undiagnosed. Tsetse flies are found in 36 African countries, putting 70 million people at risk.

Trypanosomiasis, or sleeping sickness, is a tropical disease because its vector, the tsetse fly, requires warm, humid tropical habitats. The protozoa responsible for the disease undergoes several transformations as it moves from fly to human and back again. The disease cannot be transmitted from human to human, nor from fly to fly. Instead it has adapted to take advantage of the fact that the tsetse fly requires regular blood meals to develop a life cycle that moves between hosts.

When humans are not present, other hosts will suffice. In addition to cattle and other domestic animals, game animals such as waterbuck, hertebeest, warthog and impala can also host this parasite, providing an animal reservoir for the disease. Displacement of people into rural areas, long periods of encampment and sleeping outdoors, thus have much potential to increase contact with flies that have been infected from wild or domestic animals.

Sleeping sickness is a pretty nasty business. The first sign of infection is a sore at the site of infection, at the fly bite. The sore is a localized infection and does not appear to be as serious as it, in fact, is. Depending on which species of protozoa is causing the lesion, it may remain localized for a couple of weeks to several months, after which it enters the blood stream. At this point the patient will experience periodic fever spikes. These could be mistaken for malaria, which has similar symptoms. The current explanation for these spikes is frequent changes of surface proteins on the protozoa, perhaps the result of mutation within the body. As the human immune system mobilizes in response to one set of proteins and removes protozoa displaying these, another set emerges and grows until the immune response can again recognize and overcome it. This stage of intermittent fever is called the "first stage" of sleeping sickness.

At some point (several weeks to several years depending on which protozoa species is present; there are three) the infection crosses into the central nervous system. This is the "second stage" of infection, characterized by headache, sleep disturbance, and depression, followed by mental deterioration, seizures and palsies, ultimately ending in coma followed by death in 100 percent of cases. The potentially long first stage of the disease, coupled with its close symptomatic resemblance to malaria, is the reason for screening such large numbers of the population (more than 2.4 million since 1986 by Doctors Without Borders alone).

There is no preventative medicine for this disease. Treatment depends on which of the two stages the patient has. The first stage is difficult to diagnose but relatively easy to treat with Pentamidine (for one of the species) or Suramin (for the other). Both of these drugs have possible side effects but are much less dangerous than the treatment for second stage trypanosomiasis. Treating the second stage of this disease requires a drug that can cross from the blood stream into the nervous system tissue, as the organism has already done. Only one candidate exists, Melarsoprol. An arsenic derivative, this medicine is extremely toxic, producing side effects ranging from myocarditis and renal damage to neural damage and encephalopathy. The World Health Organization estimates that 5 percent of patients die from the medication while 5 percent relapse. Doctors Without Borders puts the death rate from Melarsoprol at 5 to 20 percent. They also remark on how painful the drug is, with patients describing treatment as "fire in the veins".

A doctor who discovers sleeping sickness in its early stage is lucky, because the patient can take Pentamidine (for example) rather than resorting to stronger medicine. This drug is produced by Sanofi-aventis and distributed in collaboration with the World Health Organization. The pharmaceutical firm Sanofi-aventis described itself in 2009 as "the only industrial partner engaged in the fight against this disease".

Development of a drug like Pentamidine requires testing both its efficacy and its pathway through the body. All drugs eventually leave the body, whether through metabolism or elimination via the kidneys or liver. In order to give guidelines for doses, the rates at which metabolism and elimination   take place must be both measured and modeled. Medicines that must be absorbed into the blood stream and dispersed into tissue have two more processes that must be understood and modeled. Pentamidine is relatively simple, as it is delivered via intravenous bolus and disperses relatively rapidly compared to the rate at which it is removed from the body. So for this drug, a model can safely ignore the processes of absorption and dispersion. The drug does not appear to be metabolized, but is removed from the body through the liver or the kidneys.

Organs that remove toxic substances from the blood stream do so via semi-permeable membranes, pictured in Figure 3.1. In some cases the membranes have special "doorways" through which a molecule can move in one direction only. In other cases the molecules might just diffuse through the membrane.

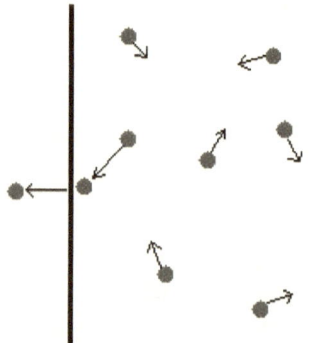

Figure 3.1: Diffusion through a membrane.

The quantity of molecules that are passing through the membrane per time unit depends on the chance of a molecule bumping into that membrane. That chance depends proportionally on the concentration of molecules present in the fluid (in this case the blood). This simple relationship is actually describing a differential equation. If $C(t)$ is the concentration of drug in the blood stream at time t, then the rate at which it is being removed is proportional to $C(t)$. That is,

$$C' = -kC$$

The negative sign indicates that the change is negative and k is some constant that depends on the particular drug and, to a lesser extent, the particular patient. Of course, this is the same equation we studied in Chapter 1 of this text with respect to population growth, except that the constant, $-k$, is a negative number. Nonetheless calculus tells us the solution:

$$C(t) = (\text{apparent initial concentration}) \times e^{-kt} = ce^{-kt}$$

So we see that the solution is a decreasing exponential function as pictured in Figure 3.2.

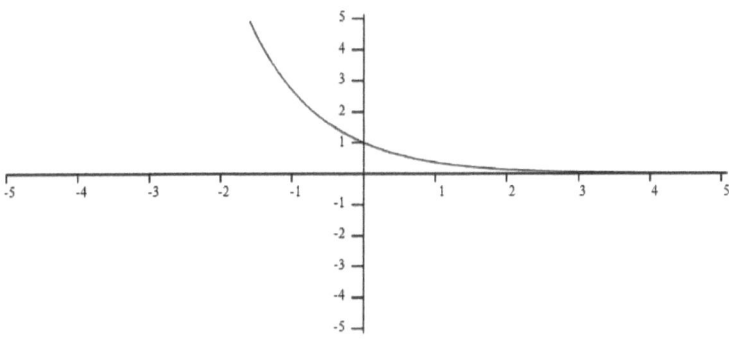

Figure 3.2: $C(t) = ce^{-kt}$

Immediately after receiving the IV bolus, the drug is being removed from the body. At some point there is only half as much left. When is this? We can set up a simple equation:

$$\text{Half as much left} = \text{the value of the solution at some time}$$

$$c/2 = ce^{-kT}$$

or

$$\frac{1}{2} = e^{-kT}$$

or

$$\ln(1/2) = -kT$$

or

$$\ln(2) = kT$$

or

$$T = \ln(2)/k$$

or

$$k = \ln(2)/T$$

Notice a couple of things.

T, called the "half-life" of the drug, does not depend on initial concentration, only on k. In the pharmaceutical literature you might not find k, but you might find T instead.

As k, the "constant of elimination", increases, the half-life decreases. This makes sense. If you increase the rate of removal, the drug should spend less time in the blood stream and drop faster. Impaired kidneys could reduce k thus increasing the half-life of the drug. If you are a doctor giving medicine to a person with only one kidney, and if the kidney is the only organ of removal for that medicine, then that person has only half the constant of elimination of a normal person. If the medicine is an IV bolus with simple elimination dynamics like this one, then you had better double the half-life stated in the literature, otherwise you might poison your patient.

What about the initial dose? C is given as a concentration of drug in the blood. Pharmacokinetics texts are careful to point out that the blood is just a proxy for the amount of drug in the entire body. Some of it is in soft tissue and unavailable to be measured. So there is a fudge factor that is merely a fraction: namely the fraction of drug visible to someone taking a blood sample. This can be measured directly in the lab. In other words, if you put 4 mg of drug in an IV bolus and administer it to the patient, wait a bit, then take a blood sample, it may appear that too little is present. In fact, it may appear (based on the model) that the patient has way too much blood, and the drug is quite dilute. For this reason, "c" in our equation above might really be somewhat less than 4. We say it is c/V, where V is the "apparent volume of blood". V varies with the drug under consideration.

According to the World Health Organization, the recommended dose of pentamidine is 4mg per kg of body weight. In order to get a certain initial concentration of drug in the body, more must be given to larger people. So a patient weighing 100kg will receive a dose of 400mg in an IV bolus. This is to be given once per day. The half-life of pentamidine is 6.4-9.4 hours. This is quite a spread. The apparent volume of pentamidine for IV bolus administration is 140 liters. We will use this number, although it was measured in studies on AIDS patients who may not truly represent a population of trypanosomiasis victims.

So for pentamidine, the equation describing blood concentration over time is this:

$$C' = -k \times C$$

And its solution for this dose looks like this:

$$C(t) = (400\text{mg}/140 \text{ liters})e^{(-(\ln(2)/6.4)t)}$$

Now with our model we can answer a few questions.

1. How much pentamidine is left in the body after 24 hours?

2. If the dose is repeated after 24 hours, how much will the initial concentration really be? (taking into account what is left in the body)

3. If the second dose is given 2 hours early or 2 hours late, how will this affect initial concentration for that dose?

4. If the half-life is really 9.4 hours instead of 6.4 (since there seems to be a discrepancy), how do the initial concentrations vary for the second dose?

High concentrations of pentamidine can cause kidney problems.
What we really don't want to see in these repeated doses is a
constant increase of initial values, driving the concentration higher
and higher over an extended time period. How many doses do
we have to give before the initial concentration becomes 10 percent
greater than what is intended (which can be inferred from the World
Health Organization instructions)?

These and other questions are all part of the cure for trypanosomiasis. Recent epidemics in various parts of Africa have reported
a prevalence of 50%, surpassing HIV/AIDS in mortality rates in
those regions. Pharmaceutical research and the models that in-
form it are a necessary part of the cure for sleeping sickness, which
also encompasses distribution strategies for medical care, the movement of populations, the biology of the tsetse fly and the ecology
that supports it. The discussion of pharmacokinetics in this text is
similar to material found in any standard pharmacology textbook.
The reader is invited to consult Atkinson *et al* (2001) and Bourne
(online source) for further information. Any reader interested in
investigating the pharmacology of treatments for tropical diseases
such as those discussed in this text can find the relevant constants
and treatment regimes in Gustafsson *et al* (1987).

Chapter 4

Okoko

Some critics will undoubtedly call this book an attempt to use Africa's history and geography as an excuse to do some math. Others will claim that we are using mathematics as an excuse to think about African issues or even politics. The truth is quite a bit more complicated than that. The models we have in our heads, whether they be of ecological systems, other cultures, or human behavior, all determine our point of view towards these realities and dictate our behavior. Nineteenth century Europeans, for example, had a model of African culture informed by the writings of Speke and other explorers, who offered a picture of an East Africa perfectly designed to justify later colonization.

The truth was again more complicated than the writings of those explorers would allow. What follows here is a brief incomplete sketch of several hundred years of African history condensed into a few paragraphs, that are included to make an important point about how people think and make decisions. The discrepancy between reality and model is an issue plaguing all disciplines. The English usage of the word, "tribe" to describe peoples of Africa nicely illustrates this difficulty. No definition of the word allows for the full range of political and cultural behavior of African tribes. For example, some historians claim that early nineteenth-century Tanzania lacked discrete, compact and identifiable tribes, whether measured by territory, language, culture or political system. Centuries of migration, mingling and contact had interwoven peoples from four different language groups and economic backgrounds; Khoison hunters, Cushitic and Nilotic herdsmen and Bantu and Cushitic farmers. The result was a diverse heritage for those living in Tanzania. Anthropologists have found many Bantu speaking tribes claim Cushitic heritage, Nilotic peoples claiming a Bantu background, etc. Amidst these peoples are more distinctive and individualistic tribes including the Ha living on the shores of Lake Tanganyika bordering western Tanzania and the Ngoni of the southern highlands.

Similar peoples populated Kenya, although there the tribal structure was more intact and local culture and custom offered a clearer partition of the population into tribal groups. The land had an eclectic mix of Cushitic herdsmen such as the Somalis in the dry Northeast, Nilotic pastoralists such as the Massai of the highlands and Bantu farmers in a more agriculturally conducive Southeast and Southwest. Here, tribes seemed to be more unified, although diffusion among different tribes was possible and the notion of "tribe" in this precolonial world differs markedly from a common western conception of the term. Even today, censuses regularly report unexplained drops or gains in certain "tribes" as their peoples change occupation or some other factor and consider themselves of new ethnicity as a result. Of course, the confusion of census takers may reflect the uncertainty with which an individual regards his or her own affiliations.

Uganda offers a political picture contrasting sharply with that of Kenya. Languages spoken in precolonial Uganda were of Nilotic and Bantu decent with the Nilotic herdsmen peopling the arid regions in the North while the Bantus farmed the South. Political organization was on a larger scale in Uganda, with single governments controlling large amounts of land. Situated on the shore of Lake Victoria, Buganda was the largest and most powerful state in Uganda by the middle of the eighteenth century and would play an important role in the country's future development. A large number of tribes inhabited Buganda as well, retaining their identity to some extent within a larger political system. The existence of the Bugandan state would dramatically alter the development of Uganda away from the experiences in both Kenya and Tanzania. The British ended up ruling much of Uganda indirectly through their Bugandan allies. The colony remained, in large part, African controlled. Kenya, on the other hand, suffered direct European rule and was politically dominated by its white immigrants for a long time. As the struggle for independence progressed, the Kenyan tribes fought the colonists in isolated camps. Kikuyus, Embus and Merus formed a guerrilla warfare group called the Mau Mau (which stood for "muzungu aende kwao mwaafrica atawale"; the white-man should go to his home country and the African must govern himself). Luos formed intellectual pressure groups intended to re-move the colonialists from power using dialogue. The two groups came together and formed a formidable force that at last made the colonialists relinquish power.

One might say that the European mindset that included a model of Africans as ignorant tribesmen organized into cultural groups in local competition with each other had its best (but not very good) fit in Kenya where tribal groups were somewhat distinct from one another, and its worst fit in Uganda where political states were large and well organized. History springs from a complicated interaction between the cultural and political reality of one group and the model of that reality in the minds of another. A dialogue must therefore ensue, that allows our mental model of a particular culture to adjust itself as further information comes to light.

The point of this historical observation is that science works *exactly the same way.* There is the reality of the organism, the ecosystem, the lake. Then there is the model we have in our heads about it. The dialogue between the two is carried out via math-ematics. We mathematize our model in order to make predictions about reality. We then test our predictions by taking measurements either in a laboratory or in the field. This is how we invite reality to take part in the conversation, either confirming our model of it or suggesting alterations. The tools of mathematics are a rich lan-guage, offering both computational and visual means for comparing our mental models of reality to the reality itself.

One way to gain insight into the making of models (mathemat-ical or otherwise) is to attempt to view the complicated system under consideration from the point of view of one of its compo-nent populations. For example, from the point of view of an algae awash in a flood of fertilizer, resources really do look unlimited and the Malthusian model accurately represents the world according to this particular organism. The enthusiasm of Europeans over the commercial prospects of colonizing Africa rings with a similar tone. Resources looked endless and the "expansionist" views of nations settling their peoples in those areas were as natural as the bloom of algae in the presence of a rich source of nutrient and suffered from a similar level of introspection.

Every ecologist knows that resources are, in fact, limited. Populations recognize limited resources both by individual behavior and also by demographic features of the population itself. Even in the presence of increased algae production, organisms higher in the food web might experience resource difficulties of other sorts, such as limited oxygen or limited habitat, which prevent them from taking full advantage of increased food. So, if we want to study an organism feeling the pinch of limited resources, we might look higher up the food chain.

The ecosystem of Lake Victoria, like most ecosystems, is based on complex interactions among species in many trophic levels. Very generally, at the bottom of the food web are producers (organisms which convert energy into a form that can be used by other organisms) such as algae and diatoms. Small fish in the food web feed on such things as phytoplankton (such as blue green algae and diatoms), zooplankton, insects, fish eggs, and mollusks as well as many others which fed on detritus (dead and decaying matter). Many of these fish with such diverse feeding habits are cichlids (*haplochromines*) and tilapia). Most of them are "feeding specialists", meaning these fish ate only one or a few different types of food. At the top of the food web were piscivorous fish such as the Bagrus catfish. Not too long ago there were hundreds of distinct cichlid species in the lake.

We have already come up with a model for an organism that is at the very bottom of the food web. Let us see if a more complex organism from the top of the food web can be similarly modeled. The catfish, *Bagrus docmac*, is one of the top predators in the Lake Victoria ecosystem. In trying to model its population changes we will start by making the same assumptions that we made for the algae population:

1) There is no immigration or emigration.

2) There is unlimited resource.

3) The catfish population increases at a constant per capita (or per catfish) rate, r. Based on these assumptions, the rate of change of the catfish population is the product of the catfish's per capita rate of population increase (r) and the population size (N):

$$\frac{dN}{dt} = rN$$

Once again we use N to represent the population size, as this is a standard notation. As with the equation for the algae population, when we solve the equation for the catfish population we find that:

$$N(t) = N_0 e^{rt}$$

If this were really the case, we would expect the population of catfish to increase exponentially.

If we look at the data for the Bagrus catfish in Lake Victoria in Figure 4.1, we will see that this is not a good model. No matter for which country we look at the data, the population of Bagrus catfish in Lake Victoria does not increase exponentially.

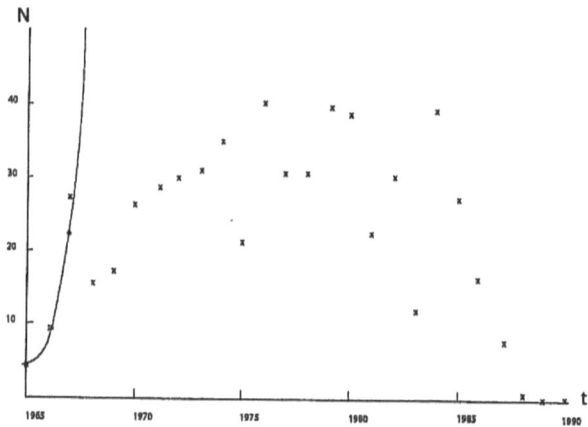

Figure 4.1: $N(t) = 4.24e^{.836t}$ and catfish data.

In the exponential model, as time increases towards infinity, the size of the catfish population also approaches infinity. If resources are not really unlimited we might want to modify our original model to include the limitations on resources and space that exist in Lake Victoria. We might assume that there is an upper limit, called a carrying capacity (K), on the number of catfish the lake can sustain under these limitations. One way to view the situation is that the catfish population with unlimited resource and space can increase at a rate proportional to the size of the population, as it does in the exponential model above. If there is an upper limit (K) to the amount of catfish that can live in Lake Victoria, then at any given time, some fraction of this limiting value represents available resources for growth. This fraction is the quotient of the difference between the carrying capacity and the number of catfish in the lake and the carrying capacity:

$$\frac{(K - N)}{K}$$

or equivalently,

$$\left(1 - \left(\frac{N}{K}\right)\right)$$

Ecologists think of this expression as the unfulfilled potential of the lake to support catfish. It may be used to estimate the effects of the existence of a carrying capacity on the rate of change of the population. How can we modify our equation for the catfish population to include this fraction? If we multiply the rate of growth by the number of catfish that can still be sustained by the lake, we get

$$\frac{dN}{dt}$$

$$= r \text{ (population) (unfulfilled potential)}$$

$$= rN(1 - (\frac{N}{K}))$$

Does this new model, called the "logistic equation", fit our assumptions about carrying capacity? The expression on the right hand side of the equation is a double proportion. That is, growth is proportional to both the amount of organism present and the resource available for growth. Double either of these and you double the rate of growth. Remove either the organism or the space available and there can be no growth. This expression is just an approximation to reality, but it seems to represent more of our assumptions than simple exponential growth.

We also know that when the carrying capacity has been reached (the population of catfish is equal to the maximum sustainable population $(N = K)$ the size of the population should stabilize. In other words, $\frac{dN}{dt}$ ought to equal zero. For our model when $N = K$,

$$\frac{dN}{dt} = rK(1 - 1) = 0$$

so our model satisfies this assumption.

We also expect the ecosystem to be unable to support more catfish than its carrying capacity. So if N becomes greater than K, the population will decrease because some members of the population will not have enough resource or space. The rate of change of the population will be negative. In our model, if $N > K$ then $(1 - (\frac{N}{K}))$ is negative. Since we defined both r and N to be positive, $\frac{dN}{dt}$ must be negative. Our model therefore appears to fit this assumption as well.

This model seems to be a reasonable one. We have set up equations that reflect our very basic assumptions about the ecological circumstances of the catfish. Of course, there might be other equations which satisfy these assumptions too. So, if our model fails to capture the properties of the data we collect, it might not be due to wrong assumptions but to a wrong choice of mathematical model. On the other hand, our model is so simple that we might suspect that fancier equations would be saying more about our situation than we really intended to say. So, by sticking with the simplest equations that express our underlying beliefs about the system, we are putting our mental picture of the catfish-in-a-lake more fully to the test. We are asking, in effect, whether our simple assumptions about limited resources are powerful enough to predict the long term behavior of the system without having to take more subtle effects into account. We will use the mathematics of the model as a mediator between our ideas about the system and the reality offered to us by the data.

If we solve the differential equation

$$\frac{dN}{dt} = rN(1 - (\frac{N}{K}))$$

for N analytically (which means finding a known function that obeys the equation, either by hand or using a symbolic computer program), we can determine a function that describes the size of the population of catfish in terms of time:

$$N(t) = \frac{KC}{(C + e^{-rt})}$$

We can choose constants r, c, and K so that we can look at this function graphically. When we do so, we get an S-shaped curve like the one shown in Figure 4.2.

This function may then be used to obtain the size of the population at any particular time or to predict long-term behavior of the population. We can try to fit a curve of this form to data, in order to deduce the values of the three constants involved and learn

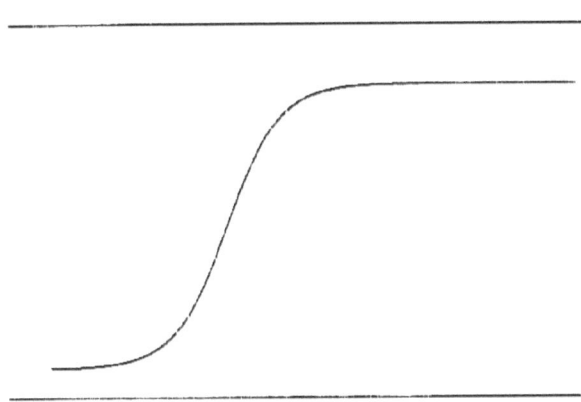

Figure 4.2: A typical solution to the logistic equation.

something about the intrinisic rate of reproduction of catfish. We could compare the shape of this family of curves to data to see how well our basic assumptions are borne out in nature.

Of course, it is a little harder to study families of curves than it is to study a single curve. When looking at ecosystems it is usually quite hard to say for sure which particular solution to a differential equation is the best stand-in for the behavior of the real system. A more cautious approach would be to make statements which are true of the set of solutions as a whole, rather than choosing a particular solution to study. If we know something about the global behavior of all, or nearly all, solutions, then we know a small error in our assignment of a particular solution to a particular situation or set of data won't result in qualitatively different predictive outcomes.

For example, we could ask whether, for all of the solutions described by our model, there are quantities of catfish for which the catfish population size is not changing. In other words, where does this model have equilibrium solutions? Once again we need to set the rate of change ($\frac{dN}{dt}$) equal to zero to determine when the population of catfish (N) is constant.

$$0 = rN(1 - \frac{N}{K}))$$

Since r is positive,

$$N = 0$$

or

$$(1 - \frac{N}{K}) = 0$$

and therefore,

$$N = 0$$

or

$$N = K$$

Fortunately, in this model we are blessed with an expression so simple that it factors completely and gives all equilibrium solutions just through algebra. Most models are not so amenable to analysis. When the model is more complicated, we resort to computational methods and display the results of our computation so that features like equilibrium points are visually obvious. Figure 4.2 clearly shows a function approaching an equilibrium as time increases. But this is only one of many possible solutions to our differential equation. We get a different solution for each starting population. In order to see if all of them have the same behavior, we can analyze a slightly different graphical output: the phase portrait.

A phase portrait is a graph where time is not one of the axis variables. For this phase graph we will plot N along the y-axis. At each point on the y-axis we can indicate by an arrow the direction the population is moving when it is that number. In other words, an arrow pointing up indicates increasing population, one pointing down indicates decreasing. The size of the arrow indicates magnitude of change. This allows us to see the relationship between the size of the population and its rate of change. The phase portrait for our model is adjacent to the vertical axis shown in Figure 4.3.

On such a graph, because t is not a variable, we do not need to have a t axis to look at the long-term behavior of solutions. Reducing the number of variables we have to plot can be useful visually. Phase graphs enable us to look at more than one solution at once,

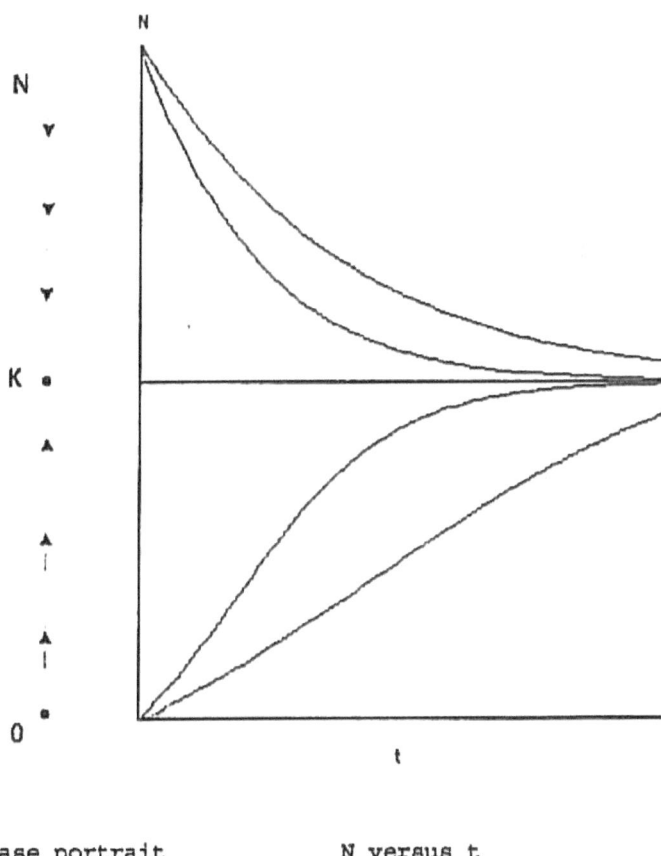

Phase portrait   N versus t

Figure 4.3: A phase graph for the logistic equation. Only the vertical axis shows the phase graph.

because each point on the graph represents a possible initial condition, and every path tangent to the arrows represents a possible
solution to the equation. Best of all, they are useful for analysis of a
differential equation that can't be solved analytically, because they
can be generated using numerical techniques on a good computer.
The curve of a solution on a phase graph that indicates the direction
of increase of time is called a trajectory. In the phase portrait in
Figure 4.3 the trajectory of any solution is a segment of a straight
line. The graph on the right hand side of Figure 4.3 shows several
solutions displayed as "time series" with time as the horizontal variable, just as we are accustomed to seeing it. On the left is the phase
portrait which only displays increase and decrease of N. We can see
all the various long term behaviors of N at once this way, although
we lose information about how fast N is increasing or decreasing.

Information about equilibrium solutions is particularly easy to
read from a phase portrait. There are two types of equilibria: stable
and unstable. Because these solutions are not changing in time,
equilibria appear as points on phase graphs, and thus they are called
equilibrium points or fixed points. If all the trajectories near a
particular equilibrium point end at that equilibrium, then we say
the equilibrium point is stable. The size of any population which is
not quite at equilibrium will eventually attain its equilibrium value
or at least remain trapped nearby. Instability on the other hand
means that a small change in the population size will cause a major
change in the system. The size of a population will not return to
nor even remain near its equilibrium value.

The knowledge of stability may be used to predict what will
happen in a system that does not start in equilibrium. Towards
which equilibrium state, if any, will the population tend? There are
equilibrium values at both $N = 0$ and $N = K$. Without solving the
equation, we can see what the trajectory of a solution that starts
near an equilibrium point will do. Will it go towards, away from, or
around the equilibrium point? If we begin with $N = K + a$, where
a is a small positive quantity, the rate of change of the catfish population $\left(\frac{dN}{dt}\right)$ will be negative as the phase graph indicates. This
means that the population (N) is decreasing. As N decreases, a
gets smaller and the rate of change of the population becomes less

negative (i.e. it gets smaller in absolute value, but remains negative). The number of catfish in the population (N) continues to decrease but the rate of decrease gets smaller and approaches zero. The trajectory approaches the equilibrium solution.

If a is a small negative quantity, the rate of change of the catfish population will now be positive and the size of the population increases. As the catfish population increases, the rate of change decreases. So, as the population (N) continues to increase from a value less than K, the derivative gets smaller and approaches zero. This trajectory also approaches the equilibrium point $N = K$.

At $N = K$ however, we reach a stable equilibrium solution. A small change in population around $N = K$ will result in the population increasing or decreasing slightly to return to the equilibrium solution. The size of a population of catfish will, according to this model, tend towards the carrying capacity. We can verify this mathematically by taking the limit as time approaches infinity of $N(t)$.

As long as r is positive (which we assert it is), as time elapses N will approach K. We follow a similar approach now for the equilibrium point at $N = 0$. We start with a trajectory that begins at $N = a$, where a must be a small positive quantity (If a were negative, we would have a negative population size which, of course, is impossible). The rate of change of the catfish population will be positive as shown on the phase graph. The population (N) increases away from zero. Thus a small change from $N_0 = 0$ produces a trajectory that tends away from the equilibrium point $N = 0$.

We say that $N = 0$ is an unstable equilibrium because a small change in catfish population will cause the population to continue to grow away from that equilibrium point (rather than return to that equilibrium). This situation is illustrated if one thinks of a system in which there are no catfish to begin with ($N_0 = 0$). There is no way for this system to start miraculously producing catfish. Both N and dN/dt stay at zero. However, if someone were to introduce even a small number of catfish, according to this model the population will grow. The population will grow large (diverging from zero) and not decrease again unless other factors are introduced. Growth will slow down as the population gets near its ceiling, K.

We can compare our graph of the catfish data to a choice of curves for $N(t)$. When we again use the data from Uganda, we get the graph shown in Figure 4.4. We see that there is still quite a bit of error in our model. One of the main reasons for this may have to do with the fact that the cichlids, the main source of food for catfish, decreased in number over this period of time. Not only is there a limited amount of resource available but that amount decreases; it is not constant as our model assumes.

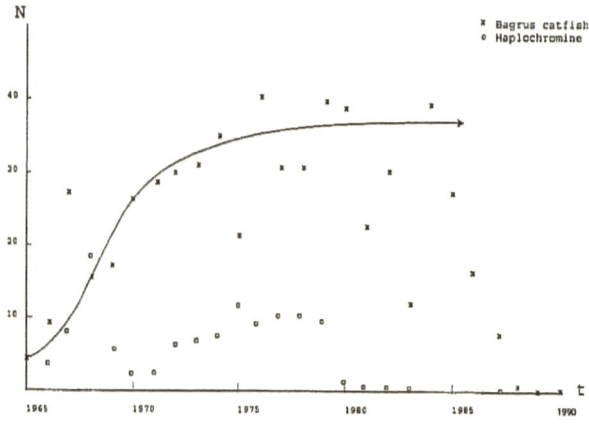

Figure 4.4: The logistic curve and the catch data.

For your consideration

Question 1:

Do you believe that the disagreement between the data on the Bagrus catfish and the logistic model is a result of a decrease in prey fish, as is claimed in this chapter? The fall-off of of the percentage of catfish caught is probably an indication of lower population, but what about the sudden increase in percent of catch? Could this be explained by other factors? Analyze these factors, decide which are likely to have most prominent effects, and then describe how you might modify the model for catfish population as a result. Catch data for Bagrus catfish, tilapia, cichlid and other fish in Lake Victoria can be found online. How does the logistic model fit with these data? How about your refined model?

Question 2:

Suppose you have two tribes living next to each other, each increasing in population according to the logistic equation. In addition, every year there is a small percentage of those in tribe A who will become a member of tribe B, and *vice-versa*. Each tribe's loss is a gain for the other. How would you model the change of population in both tribes? What equations would govern the system? What, if any, are the equilibria?

Chapter 5

Extinction

The logistic equation we just studied is a far better model of how organisms grow than exponential growth was, unless the population is at the very start of its growth pattern. Even though the fit to Bagrus catch data was rotten, the logistic model captured one very important aspect of it: not growing without bound. No species can grow without bound because of limited resources. The logistic model therefore has another advantage: it builds in the effect of limited resources explicitly, so that the mathematical consequence is causally tied to our hypothesis of what causes it. We always seek this elegance in a model. We want our hypotheses clearly expressed in the equations we create, and we want the solutions to those equations, which illustrate the consequences of our hypotheses, to match our observations of nature.

Mathematically, here is a way to think about the Bagrus data. If we compare the exponential model to the data, the error in predicting a data point from the model gets larger and larger as time goes on. The exponential function keeps increasing but the data are bounded, so the error goes to infinity with time. The logistic curve, being bounded itself, results in an error that, although large, is bounded over time. One must admit that the improvement from an infinite error to a finite one is impressive, even though the finite error may (as in this case) be very large and the fit to data is not at all convincing.

Populations that really do grow in limited environments with no other forces acting on them do indeed obey the logistic equation fairly closely. It may be hard to imagine such a situation in nature but it is easy to imagine setting one up in a laboratory. Small organisms such as bacteria or insects that have a constant source of nutrient can actually be observed to grow according to the logistic equation. In nature, more complicated things usually happen. When we try to model these complicated things, a good strategy is to go after basic qualitative observations before attempting to match numerical data. In this way we can systematically test the validity of our assumptions of causality by building them into the equations we are using, teasing out which assumptions are required for a certain qualitative phenomenon to occur. The idea of a limit to growth is an example of such an observed phenomenon (in reverse, as nobody has ever seen an example of unlimited growth). The corresponding idea of limited resources is the causal explanation built into the logistic equation.

Let us look now at another observed phenomenon: extinction. The cichlids of Lake Victoria are not a single species but many. Although related through ancestry and evolution, these species maintain their current distinctions by inhabiting different niches. Differing diets and distinct habitat preferences keep cichlid populations separate, allowing coexistence of many seemingly similar species. However, over the course of the last century many species of cichlid have gone extinct in Lake Victoria. These species, if they currently exist, live in household aquariums around the world, propagated and disbursed by fish enthusiasts. But they are gone from their birthplace.

Species can go extinct and do so regularly. Ecologists offer a variety of hypotheses when this happens. Reasons from direct human interference such as habitat loss or overfishing, to "natural" causes that may be influenced directly by humans or the larger environment, such as introduction of new species or habitat change due to weather patterns. Extinctions, though they have occurred throughout the history of life on earth, have come to be considered a bad thing because so often the cause is attributed to human interference and because the incidence of extinction appears to have increased. But it is important to note that under some circumstances, extinction is desirable. For example, if the organism is a human disease such as smallpox and the habitat for that organism is the human population itself, then extinction of the organism means eradication of a human disease. If the organism is a single infection inside a person, then extinction of the species (at least in its immediate habitat) means the disease is cured. These examples also illustrate that ecological models and medical models are closely related, the difference often being just that of context and interpretation.

Let us take the example of fishing. What does it mean for a species to be fished to extinction? The possibility of fishing a species to extinction is now discussed regularly with regard to valuable commercial varieties, such as the Atlantic cod and in Lake Victoria, the Nile Perch. Does "fishing to extinction" mean one would have to catch every last cod in the Atlantic Ocean? This seems like an improbable requirement. Once the cod population drops below a commercially viable number the fishermen won't go to the trouble to catch them. Big fishing operations catch species indiscriminately but even they will not be economically viable if fish populations drop enough. But it might be possible to fish a species to extinction by reducing the population of the species to the point where it cannot, for whatever reason, procreate faster that it dies off. Catching every individual (or mating pair) might not be necessary.

In the 1950's Warder Allee pointed out the existence of such an effect in natural populations, which has since come to be known as the Allee effect. Researchers have proposed many reasons for this effect, depending on the species under consideration. Often the difficulty of locating mates at low population densities is cited, or in some cases the need for many adults in the rearing of broods. Predator behavior patterns are also a possible justification. Any model that needs to address questions of extinction must display some kind of Allee effect at low population sizes. Ecologists sometimes refer to an "extinction threshold". Their guess is that there is some cutoff for population size. If the population were to fall below that number, it would automatically go extinct (usually for one of the reasons cited above). In terms of modeling, we would say that we expect the hypothesis (difficulty of locating mates, for example), which we would build into the equations, would lead to a conclusion (the extinction threshold, usually demonstrated graphically as a phase portrait).

Let us look once again at the logistic equation and its phase portrait in Figure 5.1.

$$X' = kX(1 - X)$$

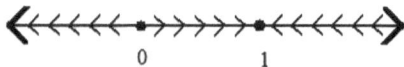

Figure 5.1: Horizontal phase portrait for the logistic equation.

It is easy to see from the phase portrait that, no matter how we adjust this organism's inherent growth rate and the carrying capacity of the environment, this model will never display an Allee effect. No matter how small the starting population is, as long as it is positive the population will rise to a stable equilibrium. This model, although excellent at reproducing observed limits to growth, is unable to reproduce observed extinctions at low population sizes. Figure 5.2 shows another way to display this constraint.

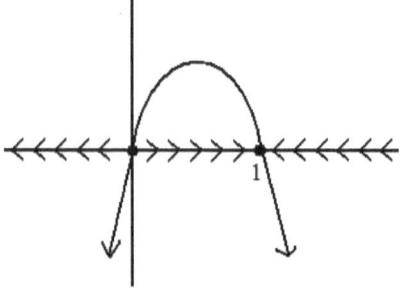

Figure 5.2: Phase portrait and graph of $aX(1 - X)$.

The function describing the differential equation for logistic growth is always positive between zero and the carrying capacity. At no point in that interval can the population drop at all. To reproduce an Allee effect we might want to see a function that looks more like the one in Figure 5.3.

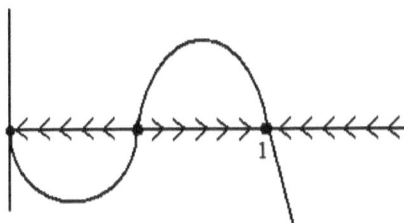

Figure 5.3: An alternative growth rule and accompanying phase portrait.

So from a modeling standpoint, the question is whether any of the reasons offered for the Allee effect as observed in nature will result in a picture like this one, which gives the desired qualitative outcome of an extinction threshold below which the population goes to zero. The extinction threshold is also an equilibrium of the system, but an unstable one. Above it the population rises to some stable value. Below it the population drops to zero.

Notice that the phase portrait displays the unstable equilibrium clearly, whereas one cannot observe it by graphing solutions against time. Unless the initial value is exactly at the critical value the solution will shoot up or down. Even if it begins at the critical value, roundoff error may be enough to push the solution towards a different equilibrium.

Now let us look at several hypotheses that might account for the existence of an Allee effect in a population. Suppose we start with a population that obeys logistic growth and add a death rate due to predation. Our equation will have this form:

$$X' = aX(1 - X) \text{ - (death due to predation)}$$

Let us consider two options for predator behavior. The predator could take a constant proportion of prey per unit time. This assumption would result in an equation like this:

$$X' = aX(1 - X) - bX$$

Solving for equilibrium values in this model yields two equilibria, one at zero and one positive value at $X = (a - b)/a$. Between these equilibria the function describing X' is positive. So the phase portrait looks like Figure 5.4.

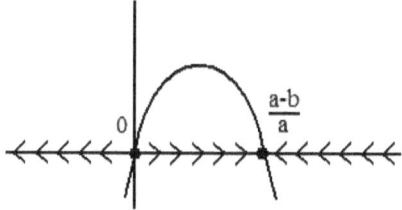

Figure 5.4: Phase portrait for $X' = aX(1 - X) - bX$

Qualitatively there is no difference between the behavior of this system and that of the original logistic equation that we modified. The population stabilizes at a value slightly below 1 and no Allee effect is present.

Alternatively one might assume the predator always takes a constant amount of prey. In that case the equation would look like this:

$$X' = aX(1 - X) - b$$

In this case there are also two equilibrium values at $X = 1/2 + / - 1/2 \times \sqrt{1 - 4b/a}$. Figure 5.5 corresponds to this function and its associated phase portrait.

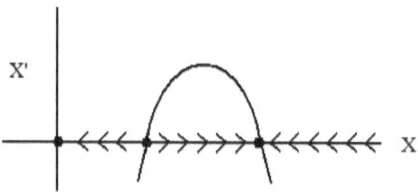

Figure 5.5: Phase portrait for $X' = aX(1 - X) - b$

The lower equilibrium is unstable and functions as an extinction threshold. Below this value the population dies off. Above it the population rises to a stable equilibrium value. As long as $a > b > 0$ we see an Allee effect in this model. As b gets small the region below the extinction threshold shrinks and the model looks more and more like the logistic equation.

There is one problem, however. In the logistic equation, $X = 0$ was a stable equilibrium. This makes intuitive sense for populations. We have lost this nice feature in the equation pictured in Figure 5.5. At $X = 0$ the growth rate is actually negative, pushing the population into negative numbers, a biologically impossible situation. The model itself becomes an unreasonable description of predator behavior as the predator attempts to consume more prey than exists.

A modeler eager to preserve the qualitative feature of the Allee effect would have to make an aesthetic decision at this point. One could use this model to study the effect with the overt recognition that it is only valid for $X > 0$. In this case it could not be used to study what might happen at very small population sizes. Alternatively a modeler could reformulate the death rate as a function that is roughly proportional to X for small values of X but approaches a constant value as X gets large. Models incorporating this feature are in the literature, as are models that consider mating difficulties and other hypotheses.

The demise of the cichlid species of Lake Victoria is often attributed to the mid 20th century introduction of the Nile Perch, a wide ranging general predator. Since the introduction of this species, a large number of cichlid species have disappeared from the lake. Models such as the ones in this chapter are consistent with the hypothesis that cichlid extinctions are linked to the introduction of this fish.

Chapter 6

It must be somethin' in the water

Fishing is only one of the activities people conduct near lakes. Sometimes the lake may be the only source of fresh water for drinking, cooking, and washing. The presence of plant material along the shores creates opportunities for basketry and furniture making, but the plants must be gathered from the water and then dried. Livestock and wild animals may use the lake as a water source. Children may play in it.

Livestock and wild animals are notorious for dropping feces into their drinking holes, but humans also deposit fecal matter in bodies of water. Sometimes the sewage systems of an entire community may drain into a large body of water and other times the droppings are just the result of accumulated individual actions. Fecal matter is certainly fertilizer and has its uses if managed well. But in areas around tropical freshwater lakes feces have an additional function.

The schistosome is a flatworm that has evolved to take advantage of the tropical paradise provided by warm lakes and ponds, the presence of mammals and their feces, and a few species of freshwater snails. The worm itself lives on red blood cells of a human or animal, settling only inside the blood vessels of the intestines or bladder. There is both a male and female form, which link together to form a permanent pair. They live about 5 years, producing 200 to 2000 eggs per day during that period, although some species can live up to 20 years inside a human host.

The eggs find their way into many organs, becoming trapped inside the body but never hatching, as they can hatch only when released into fresh water. The constant release of eggs creates chronic but not necessarily deadly disease for the host, including abdominal pain, bloody diarrhea, cough, fatigue, fever, and enlarged spleen or liver. Continued infection may cause permanent damage to organs, attributed largely to the body's own immune response to the eggs. In children the disease can interfere with growth patterns and cognitive functions. A form of bladder cancer in adults has been linked to schistosomiasis, which is 32 times as frequent in some regions of Africa where the parasite is endemic as it is in the U.S.

Some eggs find their way into urine and feces, leave the human body, and wait for contact with fresh water, at which point they open releasing a free swimming larval form, the *miracidium*. The *miracidium* enters a freshwater snail through its foot, passes through several more life stages during which it multiplies extensively, and is released back into the water in a different form, the *cercariae*. These are very mobile and responsive to water turbulence and chemical triggers. They attach themselves to the skin of humans or other animals that have entered the water, and eventually penetrate the skin. Changing form again, the parasite eventually develops into a male or female flatworm and these form pairs, producing eggs and completing the life cycle of the organism.

The World Health Organization estimates that, worldwide, over 200 million people are infected with schistosomiasis, of which 120,000 show symptoms and 20,000 suffer severe damage. Although not a deadly disease, it has serious consequences for children and does economic damage on a grand scale. It is spread by human migration and by changes in habitat that may create new water reservoirs or changes in climate. Many people who have this disease are unaware that they carry it, thus becoming unwitting vectors. Additionally, many are unaware of how it is transmitted through water contact. This disease is easy to treat, and so the World Health Organization recommends handling it regionally rather than on an individual basis. If the disease is endemic to an area, the strategy is to treat everyone in that area on a regular basis, without bothering to diagnose the disease in every individual. The result of such a strategy is not eradication of the disease, but merely morbidity control, attempting to keep cases from progressing to an advanced stage. Education of the population at risk is an accompanying strategy, as understanding the source of the disease leads to better sanitary practices and an overall reduction in new cases.

Until the 1970's the strategy of regional treatment would have been undesirable, as early treatments were as dangerous as the disease itself. Now the main treatment is a single dose of Praziquantel, which has side effects that are attributed to the death of the parasite burden rather than the medication. That is, the larger the parasite burden, the worse the side effects are. Medicine is given orally in solution. A large fraction of the drug is metabolized, never reaching the bloodstream. One could circumvent this issue by giving an IV bolus. However the treatment strategy requires treating an entire regional population more or less at once. Medicine taken orally requires no equipment and no trained health care worker. It is easier and cheaper to increase the dose taken orally than to administer shots to a large number of people.

One problem in pharmacology is figuring out how to predict blood concentrations of medicines that are taken orally and require time before some fraction of them reaches the blood. Praziquantel is an excellent example of such a drug, because the key to its success lies in deploying large quantities of single dose oral medications that do the job right on the first try.

To model this drug we need to consider two quantities: the amount of drug in the G.I. tract and the amount in the bloodstream. If the medicine were a pill we would have to worry about the time it takes to dissolve, but with a liquid medicine we can assume it just goes right into the GI tract.

The diagram shown in Figure 6.1 is called a "box model" and it depicts a compartment or box for each of the quantities that is changing in time. Most models start with a pictorial representation like this one. The trophic levels pictured in chapter 1 are an example of a box model that could lead to equations that govern a simple ecosystem. In this model we are only looking at the quantity of drug in the GI tract and the concentration of drug in the blood. There

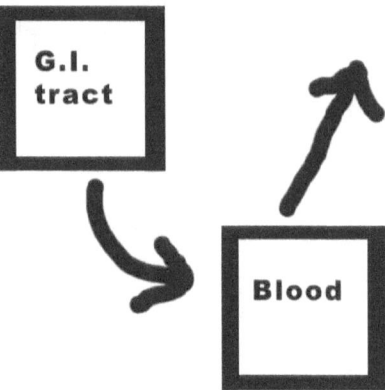

Figure 6.1: A box model for a pair of equations like those for Praziquantel.

is one box for each equation to be derived. The arrows between the boxes represent relationships between quantities that may change with time. The arrow from the GI compartment to the blood compartment represents the diffusion of drug from the GI tract into the plasma. The arrow leaving the blood compartment represents elimination of the drug by the body (via a variety of paths depending on the specific drug). The number of arrows leaving and entering a box tell us how many terms each differential equation will have. As you read the description of the equations for this system, notice how the various terms correspond to the box model.

The contents of the intestinal tract are not really liquid, so the amount of drug in it is expressed as a weight rather than a concentration. The way it reaches the blood stream is via a semipermeable membrane, similar to the situation discussed in chapter 3. The amount passing from gut to blood is proportional to the amount in the gut, for the same reasons as that previous situation. Letting $X_g(t)$ be the amount of drug in the GI tract gives us the simple equation:

$$X'_g = -k_a X_g$$

"k_a", or "constant of absorption" is a rate constant associated to this particular drug, and is experimentally determined.

If we integrate this equation we find solutions of the form:

$$X_g(t) = be^{-k_a t}$$

A decreasing exponential function with initial value b. In a perfect world "b" would be the dose of medicine we gave the patient. However Praziquantel and many other drugs have the property that only a fraction of the initial dose will ever enter the GI tract, as the rest is metabolized or otherwise unavailable. So we get a solution of the form:

$$X_g(t) = F(\text{Dose})e^{-k_a t}$$

F is called the "bioavailability" of the drug.

Now let us look at the blood concentration at time t, which we will call $C(t)$. It is the result of two processes: gain in drug from the GI tract and loss through removal by kidneys (or some other path). Both of these processes happen via membranes and obey simple proportional rate laws.

At a given time we can split the rate of change of C into two parts, rate of entry and rate of removal, and sum these. The rate of entry of drug into the blood from the GI tract is just the same as the rate at which it leaves the GI tract, except that we must convert units from weight (milligrams) to concentration (milligrams per liter). To do this we will use not the actual volume of blood in the body, but the "apparent volume" discussed in chapter 3. To summarize,

Rate of entry of drug from GI tract into blood $= X_g'/V = (k_a/V)X_g$

The rate of removal of the drug from the blood is just proportional to the blood concentration, C. We will use "k_{el}" or "constant of elimination" for the rate constant. Thus,

Rate of removal of drug from blood $= -k_{el}C$

The total change in blood concentration, C', is a result of these two processes. So

$$C' = (k_a/V)X_g - k_{el}C$$

This gives a system of two ordinary differential equations that describe the process in the box diagram above.

$$X'_g = -k_a X_g$$
$$C' = (k_a/V)X_g - k_{el}C$$

This system of equations can be solved explicitly. In fact, we already solved the first of the two. Most systems of equations that arise in biology and medicine can't be solved so nicely and require computer approximations and other, more sophisticated, mathematical techniques to study (hence this book). But this one can be, and there are several ways to do so.

We solved the first equation by separation of variables to get

$$X_g = F(\text{Dose})e^{-k_a t}$$

If the second equation were just $C' = -k_{el}C$, we could easily solve it to get $C(t) = ce^{-k_{el}t}$, but unfortunately this doesn't work. But a student who has taken a good course in ordinary differential equations would learn to make an intelligent guess in this situation. Such a student would guess the answer is some combination of the two exponential functions.

$$\text{GUESS: } C(t) = Ae^{-k_a t} + Be^{-k_{el}t}$$

Such a guess has a consequence when taking the derivative:

$$\text{Consequence of GUESS: } C'(t) = -k_a Ae^{-k_a t} - k_{el}Be^{-k_{el}t}$$

Then this brilliant student would plug this guess into the differential equation for C,

$$C' = (k_a/V)X_g - k_{el}C$$

using the "Consequence of GUESS" for the left hand side and plugging the known solution for X_g and GUESS into the right hand side. A little algebra then shows that, in order for our guess to be a solution, B can be anything but the constant A can only be one thing.

$$A = (k_a/V)(F(\text{Dose})/(k_{el} - k_a))$$

This gives a family of solutions of the form:

$$C(t) = (k_a/V)(F(\text{Dose})/(k_{el} - k_a))e^{-k_a t} + Be^{-k_{el} t}$$

Since we are only going to give one dose of Praziquantel, at time zero there will be no drug in the body. That is, $C(0) = 0$. This constraint determines what B will be. If we set $t = 0$ in the above equation, what must B be in order for the two terms to sum to zero? It has to be the negative of the coefficient of $e^{-k_a t}$. So we get a solution of the form

$$C(t) = (k_a/V)(F(\text{Dose})/(k_{el} - k_a))(e^{-k_a t} - e^{-k_{el} t})$$

All of the constants in this expression have to be determined from experiments (except the dose, of course). It is hard to measure all of these constants directly, but one experiment found in the literature shows that for an initial dose of 1800 mg in water, these constants are obtained:

$$C_{max}(water) = .637 mg/l$$

$$T_{max}(water) = 1.77 hours$$

$$k_{el}(water) = .45$$

C_{max} is the maximum concentration the drug achieves in the blood, and T_{max} is the time at which that concentration is achieved. These are easy to measure. k_{el} can be measured by using the model in chapter 2, giving an IV bolus and taking blood samples.

Maximum concentration of drug is going to occur when $C' = 0$, as we all learned in calculus. Setting $C' = 0$ and solving for Tmax gives:

$$T_{max} = (\ln(k_a) - \ln(k_{el}))/(k_a - k_{el})$$

Yielding the equation

$$(lnx - ln(.45))/(x - .45) = 1.77$$

This should be solved numerically. Once we have k_a, $Cmax$ is the value of C at $Tmax$. Plugging all the known constants and T_{max} into the equation for C will give us a value for F/V. Thus we can retrieve most of the pertinent constants from the data given by experiment. This exercise is left to the reader.

The difficulty of obtaining some of these constants experimentally is evident from the fact that different values of them appear in the literature. In particular, the *Handbook drugs for tropical parasitic infections*, (Gustafsson *et al*, 1987), gives substantially different constants for Praziquantel than do the the researchers who published the above values. As with most medicines, values for the relevant constants may also be found online.

Figure 6.2 shows a picture of a typical solution to a pair of equations like those for Praziquantel.

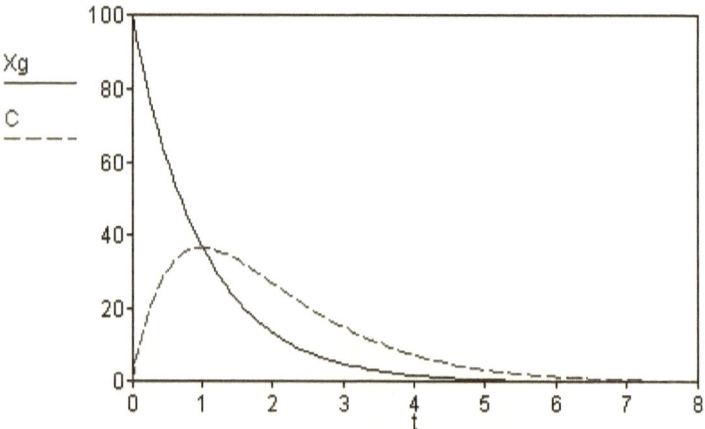

Figure 6.2: A Typical solution to a pair of equations like those for Praziquantel.

We can see that the concentration of drug in the blood rises from zero to a peak and then declines. The efficacy of a single dose will depend on how high it rises, but also on how long it stays above some minimally effective level. Sometimes it is necessary to give a fairly large dose in order to keep the level above the minimum for long enough to work. An alternative strategy (which we use for most medicines) is to give repeated smaller doses over a longer period of time, keeping the blood concentration above the minimum but much lower than the maximum that would occur with a single large dose. Many variations on these strategies can be studied with the simple two-compartment model we just studied.

The strategy that the World Health Organization recommends for controlling schistosomiasis is unusual in targeting whole populations rather than individual cases. To understand their recommendation requires more than just pharmacokinetics, which is the word for the kind of model we just developed. It also requires modeling the epidemiology of a complex disease with multiple species as vectors. In the case of poor areas, the cost is also a factor. The cost of Praziquantel is about eight cents for a 600 mg dose (and the patient takes 3 of them at once). Areas treated according to this strategy remain disease-free for 18-24 months on average, with some areas remaining disease free for up to five years. Predicting an outcome of this sort in advance is the point of a good model.

Schistosomiasis comes from a number of different species of schistosome in different parts of the world. Different types of schistosomiasis are treated often with the same, or similar medication. Differences in the parasite cause differences in the effectiveness of the same treatment, but some of the concepts are similar throughout. For additional information on Schistosomiasis and scientific studies of it treatment with Praziquantel see Homeida *et al* (1988), Belizario *et al* (2008), Frohberg & Schulze (1981), Guisse *et al* (1997), Cioli & Pica-Mattoccia (2003), Mandour *et al* (1990) and Ming-Gang (2005). Some studies suggest that the administration of Praziquantel is more effective against Schistosoma when co administered with food. For articles that explore this idea, see Castro *et al* (2002), Castro *et al* (2000), and Giorgi *et al* (2003).

For your consideration

Question 1:

What are k_a and F/V for Praziquantel, based on the data given in the chapter?

The researchers (ref) also ran tests by giving the same dose of Praziquantel, only with grapefruit juice instead of water. They obtained different results:

$Cmax$ (grapefruit juice) 1.038 mg/l

$Tmax$ (juice) 2.10 hours

k_{el} (juice) .44

What are k_a and F/V for this set of data?

The researchers concluded that giving Praziquantel with grapefruit juice increases its bioavailability. Do you agree? How would you argue this?

Question 2:

Some suggest that, because the stuff tastes horrible and gives you stomach cramps, it might be better to cut the dose in half and give two doses in one day. Use the computer to model this scenario, (for Praziquantel in water), being very careful about initial conditions for the second dose. What is the maximum blood concentration that occurs if the medicine is given in this way? Is it a good idea? What if it were given with grapefruit juice? To see what will happen you need to run the system of equations on a computer. For this you will need the value of V or F (so you can figure out the other one). These are in the literature (try searching for the Handbook of Tropical Medicine or the World Health Organization or on the name of the drug itself).

Question 3:

The World Health Organization states that 80 percent of Praziquantel in the intestine is absorbed but due to metabolization only small amounts enter circulation. Using the data from above experiments, what V do you get if the bioavailability is 80 percent?

Question 4:

WHO also gives Praziquantel a half-life of .8-1.5 hours. If Praziquantel were given as IV bolus with the constant of elimination as given above, what would the half life be? How much do the various sources for these constants vary?

Chapter 7

Developing Africa

A perusal of the publications of the World Health Organization reveals the effects of poverty and ignorance on the health of a region. Ignorance of how a disease works can contribute to its spread. Poverty adds an economic factor to medicine because methods of detection, preventions and cures must be inexpensive to administer on a large scale. The work of the World Health Organization and Doctors Without Borders often must include establishing an infrastructure for delivering care.

Rural areas have special problems because of distance of travel and endemic disease and insect vectors. Rural areas also have economic constraints because of limited resources for agriculture. So a combination of factors encourages people to migrate to cities, where they perceive their chances to be better. Increased population goes hand-in-hand with industrialization. A large concentration of people provides the workforce needed to establish factories as well as the pressing need for employment. Factories and other enterprises with jobs to offer draw people in from the countryside. This process is at work in Kenya, which now has a variety of industries from food processing to the manufacture of glass and chemicals and the assembly of vehicles. Slowly, industry and manufacture are contributing an increasing chunk of the gross domestic product. Nairobi now has its own stock exchange, listing (as of this writing) 18 industries ranging from mining to sugar refineries.

As wealth builds and goods accumulate, a region may experience new environmental hazards or diseases. In fact, we might well find regions experiencing side effects of industrialization that we recognize from earlier experiences in countries such as the U.S. The industrializing of Africa gives human beings a splendid chance to repeat their mistakes. For example, even though leaded gasoline has not been used for a while in the U.S. because of the known effects of airborne lead pollution, only recently have African countries begun to go lead-free. The decision to abolish lead from fuel is not completely a case of a belated recognition of known dangers. Rather, it waited upon the development of the effects of widespread exposure: high levels of lead in the blood of many citizens and the appearance of symptoms, especially among children.

Urban children in Nigeria, South Africa, and other sites in Africa showed elevated levels of lead in the blood during studies in the late 1990's. Researchers argued that all of Africa needed to pay attention to this problem, pointing out that a major source of lead is fuels that, in burning, release lead into the air. Air circulates, carrying lead with it to all parts of the continent regardless of local restrictions and laws. The lead passes through lung membranes directly into the plasma, with children having a higher rate of uptake. By 2004 50% of all gasoline in sub-Sahara Africa was unleaded, and Kenya announced it would phase out leaded gasoline by 2006. As an exporter of fuel to the rest of Africa, Kenya would improve the situation throughout the continent.

Leaded fuel is not the only source of airborne lead, however. Lead in paints and other products usually leach into the soil, but can become airborne if these products are burned. Garbage incineration is one way to handle the large amount of waste generated by a city such as Nairobi.

It is 2007. Nairobi's 4.5 million inhabitants produce a prodigious amount of garbage and it has to go somewhere. About 5 miles away is the town of Dandora, which features a 30 acre quarry that Nairobi has been using for a dump, depositing 2,000 metric tons of garbage each day until the former quarry is no longer a hole, but a mountain. Any kind of garbage is welcome at Dandora and to manage the incoming deluge, the garbage is burned.

Dandora is a poor area and the dump is a source of potential profit. People, including children, scour the dump for food, reusable materials, and anything they might be able to sell. As they search the dump, these people breathe the fumes from the burning garbage. So in 2007 the United Nations Environment Program, which is headquartered in Nairobi, commissioned a study of the children of Dandora. The study tested 328 children 2-18 years of age who lived near the dump. Over half of them had blood levels of lead in excess of 10 micrograms per deciliter, the internationally accepted "action" level for this poison. Fifty percent also had low hemoglobin levels and 30 percent were anemic, two symptoms of lead poisoning. Lead poisoning also eventually affects the nervous system and brain.

The study also tested soil samples in the area. 42 percent of the samples had lead levels that were 10 times higher than what is considered normal. In addition, the soil was contaminated by very high levels of mercury and cadmium. The Nairobi river runs by the dump, so any soil contaminants will eventually leach into it. Water downstream is for crops and residential use. The study was careful to point out that Dandora is not unique. Many cities burn garbage.

Lead poisoning does not produce noticeable symptoms until the blood concentration is fairly high, at which point the victim may experience nausea, fatigue, headache and general weakness. Children may develop learning disabilities, be lower in IQ and smaller in size as a result of prolonged exposure. Advanced neurological symptoms include seizure and coma.

There is no easy cure for lead poisoning. Even if a person is removed from all sources of lead, the level of lead in the blood will remain high for a prolonged period. To understand why this is the case we must look at the pharmacokinetics associated to a "dose" of lead. The way toxic substances pass through the body is sometimes called "toxicokinetics", although the approach is identical to the study of any drug.

There are two ways in which airborne lead poisoning differs from the pharmacokinetics models we already studied. One is that the lead arrives as a steady ongoing input rather than as an initial dose. The second difference is that the body stores lead in the bones. So there are two important compartments to consider: the blood plasma and the bone tissue. Lead enters the blood through capillaries in the lungs by diffusion across a membrane, is exchanged between the plasma and bone tissue in both directions, and exits the system from the plasma via the bladder. It is also stored in soft tissue, but for a relatively short amount of time. We will ignore this compartment for now but some models include it. The resulting box model is in Figure 7.1.

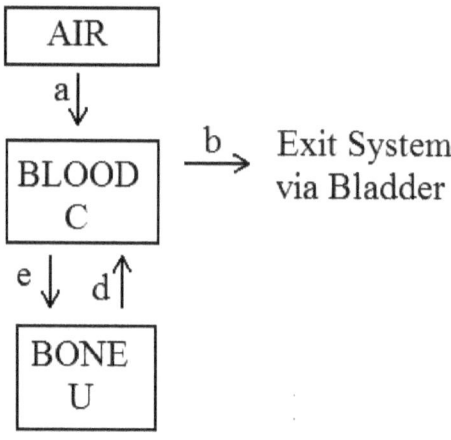

Figure 7.1: Box diagram for lead in the body.

The box model tells us that we will might need three differential equations for each of air, blood and bone. However, the amount of lead in the air is relatively constant and all we really care about is the arrow representing uptake of lead from the air. So we can leave out the first box. The differential equation for lead concentration in the blood should have four terms according to this picture, corresponding to the four arrows leaving or entering the box. Similarly, the equation for the amount of lead in bone should have two terms.

Movement of lead between compartments has the same basic kinetics as across semipermeable membranes. Rates are just proportional to concentrations or quantities inside the source compartment. However, as in the case with two compartment pharmacokinetics, the units used to measure lead concentration in blood are different from the units used to measure amount of lead stored in the bone. Thus the constants representing transfer of lead from blood to bone are different in the two equations. That is, the constants d and e represent the same transfer of lead, but in different units in each of the compartments. So we get a basic system of equations that looks like this:

C' = change of plasma concentration
= (environmental exposure) - (urine loss)
- (loss to bone) + (gain from bone)
= a - b$\times C$ - $d_1 \times C + e_1 \times U$

U' = change of lead in bone
= (gain from plasma) - (loss to plasma)
= $d_2 \times C - e_2 \times U$

If the bone could not store any lead, we would have only the simple equation:

$$C' = a - b \times C$$

No matter what the starting level of lead in the blood, after a while C stabilizes at an equilibrium value. At equilibrium, the change C' is zero, which (using the previous equation) means that $C = a/b$.

If we include the role of the bone, we still get an equilibrium where both C' and U' are zero. Solving this pair of equations gives us a solution that tells an important part of the story of lead poisoning. The bones pick up lead rapidly and lose it slowly, especially in children. That is, the constant d is large relative to the constant e. In fact, the equilibrium constant of lead in the bones is so large that you would be unlikely to reach it in a lifetime, even with persistent exposure. So although the equilibrium plasma concentration is reached fairly quickly, the lead in the bones continues to increase. Figure 7.2 depicts the resulting time series.

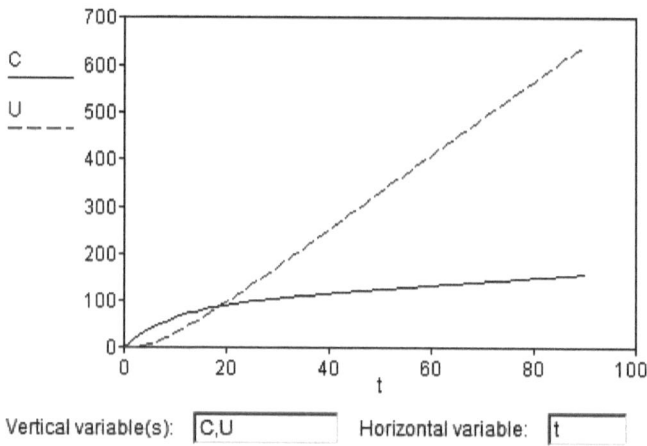

Figure 7.2: Lead accumulation in bone. Constants taken from the literature.

What happens when the source of lead is removed? That means that the constant "a" is now zero. Eventually C will also be zero, but how long will it take? This depends on how much lead has been stored in the bones. Figure 7.3 shows what happens if we take the ending values of the last graph ($C = 200$, $U = 1060$), and use these as starting values in the same system with $a = 0$.

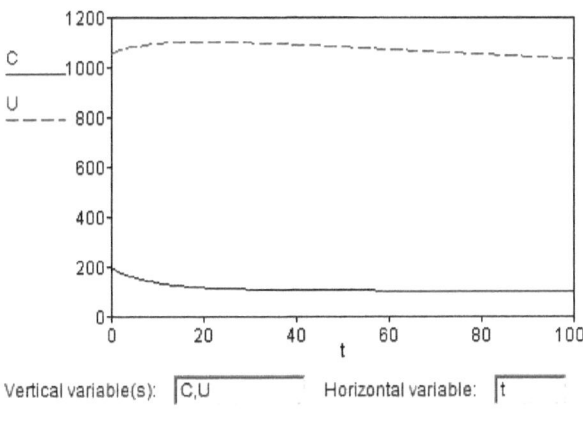

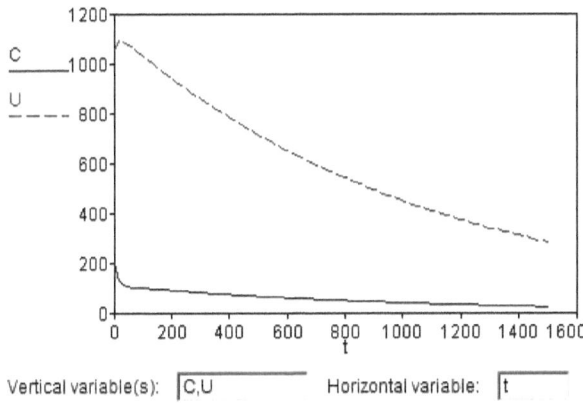

Figure 7.3: Lead elimination in a clean environment over short and long time frames.

Notice how long it takes for the lead to leave the system. The medical intervention for extreme lead poisoning (and other heavy metal poisoning) is called "chelation" and involves chemically removing lead from the blood. The patient undergoes treatment for 10-12 weeks during which an IV drip administers chemicals that bind to lead in the blood, increasing the rate at which the kidneys are able to remove it. There are unpleasant side effects to the treatment. The rate at which the removal takes place just depends linearly on the lead level in the blood. So a patient undergoing this treatment (in the absence of further exposure) has a system behaving like this:

C' = change of plasma concentration
=- (loss due to chelation) - (urine loss)
- (loss to bone) + (gain from bone)
$= - f \times C - b \times C - d_1 \times C + e_1 \times U$

U' = change of lead in bone
= (gain from plasma) - (loss to plasma)
$= d_2 \times C - e_2 \times U$

As you can see, this will increase the rate at which lead leaves the plasma but it does not increase the rate at which it leaves the bone. Figure 7.4 shows the solution of these equations with $f = 7b$, octupling the loss from plasma, compared to the previous solution with no chelation.

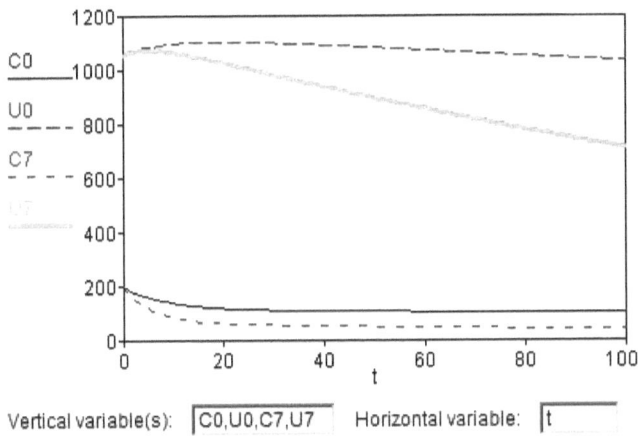

Figure 7.4: C0 and U0 correspond to no chelation; C7 and U7 correspond to chelation with $f = 7$.

As you can see, it still takes a while before the lead leaves the system. Duration of exposure is a key aspect of lead poisoning, in that the serious damage done to the nervous tissue is done over time. The bones play a major role in prolonging lead levels in the blood long after exposure has been reduced.

Airborne lead is not the only source of environmental lead in urban Africa. Soil levels and water levels are sources of exposure, as well as occupational exposure. Some diseases, such as HIV, increase susceptibility to lead poisoning. In addition other heavy metals such as mercury and cadmium create hazards. All of these environmental toxins are related to the industrialization that promises so much for Africa's economic growth, at the same time creating potential health hazards for the citizens who desire that growth.

Lead accumulation is not only a problem in the Lake Victoria region, but has presented challenges worldwide. Mathematical modeling of leads interaction with the human body is the same everywhere, so though lead studies will account for different inputs of lead worldwide, they still present useful information about the process. For more information on Lead in papers with data, and information, but no modeling, see Hu *et al* (2007) and Tsaih *et al*

(1999). For a paper including modeling see Leggett (1993).

Lead is not the only heavy metal that poses health risks to life. Mercury, also presents similar problems. The bioaccumulation of mercury in aquatic organisms, including has been studied extensively. The Daphnia, being low on the food chain presents a starting point for mercury accumulation up the food chain. For information about mercury and Daphnia with modeling see both papers by Tsui& Wang, (2004).

Because bioaccumulation is prevalent also outside of the Lake Victoria region, studies of other lakes, including Lake Erie, and Lake Murray present useful information. See Bowles *et al*, (2001).

For your consideration

Question 1:

The literature gives constants for tranfer rates of lead between blood and bone. Build a model with these constants and run it numerically on the computer. The accepted "action level" for lead poisoning 10 micrograms per deciliter of plasma. What does the exposure rate have to be to arrive at this equilibrium concentration?

Two children in the Dandora sample had blood concentrations of around 30 micrograms per deciliter. If these children are removed from this environment and placed in an environment with "normal" exposure rate, how long will it be before the level of lead in their plasma falls below 10 micrograms per deciliter?

Question 2:

We can extend the model to include transfer of lead between plasma and soft tissue. The literature has constants and models for this situation also. Build a model with this compartment included. Does it differ substantially in its response to the exposure level and recovery times you obtained for question 1?

Question 3:

Typically chelation is administered for a period of several hours at intervals throughout treatment, which is different from the outcome pictured here. After a single treatment the plasma lead level is seen to drop and then rebound due to release of more lead from the bone. Do the models in questions 1 and 2 predict this? How does the time frame of your models compare to what is seen experimentally?

Chapter 8

Fulu

In 1858, British explorer John Speke described the fishing he saw on Lake Victoria:

"There are very few canoes about here, and those are of miserable construction, and only fitted for the purpose they turn them to-catching fish close to the shore. The paddle the fishermen use is a sort of mongrel between a spade and a shovel. The fact of there being no boats of any size here, must be attributed to the want of material for constructing them. On the route from Kazé there are no trees of any girth, save the calabash, the wood of which is too soft for boat-building".

Over a century later this situation still had not changed much. In 1979, only two percent of the canoes on Lake Victoria had outboard motors. Many of the fishermen in the region do not have the capability to fish in the deeper parts of the lake and without sophisticated equipment they can not keep fish fresh if they stray far from the shore.

As of this writing, more than thirty million people in the three countries bordering Lake Victoria (Uganda, Kenya, and Tanzania) are in some way dependant on the lake for their livelihood. These people not only include fishermen but many people directly and indirectly associated with the fishing industry. Without the fishing industry, fish sellers and transporters of fish, canoe builders, net-makers, and people who repair canoes and nets would not have work. Further, owners of small shops and hotels, in addition to construction workers and railway workers, benefit greatly from the large population centered around the lake. Many additional people prosper from the reinvestment of money earned from the lake back in local businesses.

The fishing industry began to experience changes that occurred as early as the 1900s when British brought improved fishing equipment to the local fishermen. Traditionally fishermen had used traps, weirs, baskets, and spears. Flax gill nets were introduced in 1905, non-selective beach seines in the early 1920s, and synthetic fiber gill nets in 1952.

In 1908 a railway was completed from Mombasa (on the coast of East Africa) to Lake Victoria. While it took explorer John Speke nine months and an average missionary six months to reach Lake Victoria without the benefit of a railroad, after 1908 the trip only took 2.5 days. The railway facilitated the creation of a previously impractical export industry. This, in addition to a tremendous increase in population in the new urban centers of the region, led to an increased demand for fish which, coupled with the use of better fishing equipment, led to overfishing, decreasing the amount of fish in the lake. As catch size decreased, fishermen were left with no choice other than to decrease mesh size on their nets which further decreased the fish populations. In the early 1950's fishing effort nearly doubled while yield only increased by about 10 percent. The smaller mesh size meant that fishermen caught smaller fish including a significant number of juveniles. Catching young fish made it even more difficult for the populations to replenish themselves. In the early 1970's the commercial catch was as much as 35 percent immature, and some concluded that it was likely that some of the larger species of Haplochromis had almost been eliminated by overfishing.

Cichlid fish had once served as the main catch for local fishermen along with catfish, carp, and lungfish. The two endemic tilapia species *Oreochromis esculenta* and *O. variabilis* which had previously been important to the fishing industry were almost completely wiped out by the early 1950s. Although the Lake Victoria fishery should be a renewable resource, it has been exploited to the point that some feel it may not ever be able to return to its original level of productivity.

The cichlid population was in serious decline from overfishing during the same period of time as the algae bloom we investigated in Chapter 1, resulting in a major change for the lake. When the base of the food chain experiences rapid growth (as the algae did) one would expect a corresponding growth in the population of fish feeding on the algae, which include the cichlids. The extra biomass generated by the algae would then be transferred to a biomass of certain types of cichlids, which would then increase the biomass of their predators, and so on up the food chain. When the intermediate link of algae-eaters is removed or drastically reduced (as it was in this case by overfishing), the algae continue to bloom unabated. This has two consequences. First, the density of algae makes it difficult for light to reach the depths of the lake, thereby concentrating the algal and other plant biomass near the surface. Because plants produce oxygen, their concentration near the surface results in less oxygen in the depths of the lake. In fact, lake waters tend to separate into layers by temperature, making it difficult to mix oxygen produced at the surface with the deep waters. So, by their sheer numbers, the algae can change the chemistry of a large body of water. The second effect is the result of the death of a portion of the algae and other plant populations. Because nobody is eating these plants a certain proportion of them die and drop to the bottom of the lake. The decomposition of dead material by bacteria is a process which also requires oxygen, thus further diminishing the supply of oxygen in the depths of the lake. This deoxygenation forces species to move to the shallower regions of the lake where oxygen is more plentiful. Here, though, they were more likely to be caught in fishermen's nets or by predators, thus furthering the decline of the cichlid population.

The British started to stock the lake when they realized that fish populations were declining. Officials introduced the perch *Cyprinus carpio* whose introduction does not appear to have been successful and four species of tilapia (*Tilapia zillii, T. rendalli, Oreochromis niloticus* and *O. leucostictus*). The only tilapia to take hold so far is a plankton eater, the Nile tilapia (*O. niloticus*), although the other tilapia are present in small numbers. The Nile tilapia is very similar to the endemic species but it can live in a greater variety of habitats, grow at a faster rate, and therefore eats a wider range of foods.

Some British officials also wanted to introduce a large predator such as the Nile perch (*Lates niloticus*) into the lake. Cichlid and some of the other native species are small and somewhat bony. The British felt that a larger, meatier fish would be more pleasant eating, particularly for the restaurant table. Additionally, Europeans enjoyed the perch as a sporting fish because of its large size. The Nile perch, sometimes referred to as the "elephant of the water", can grow to be five or six feet long and weigh up to 135 pounds.

Although most ecologists were opposed to the introduction of the Nile perch because there was no natural predator for it, by 1954 some of this nonendemic species had somehow gotten into the lake. It is possible that floods allowed some of the fish to enter from nearby bodies of water. It is more likely, though, that some perch were intentionally introduced into Lake Victoria. Since it was already there, Ugandan officials decided to complete the process by stocking the lake with Nile perch from Lake Albert.

The Nile perch is a large predatory fish of high commercial and recreational value belonging to the family *latidae*, order Perciforms. It is a large mouthed fish, greenish or brownish above, silvery below and generally attains length of 1.8m and 140 kilograms. A length of 3 meters and a weight of about 200 kilograms has been recorded. *Lates niloticus* has an elongated, protruding lower jaw, rounded tail and two dorsal fins.

Lates niloticus has two patterns of occurrence, that is endemic and non-endemic. It is endemic in Lakes Albert, Turkana, Chad and Lakes Sharma and Abaya in Ethiopia. In the river systems it is found in the Volta and Nile. In non-endemic environment it occurs in Lakes Kyoga and Victoria, and Kabakas Lake in Kampala. Though it is considered to be non endemic in Lake Victoria, archeological records indicate that some fossils were found in Rusinga dating back to the Miocene (approximately 25 million years ago). Some fossils have also been found in Lake Edward dating back to Pleistocene period (approximately 35,000 years ago), showing that Nile perch could have been in existence in these Lakes before the recent introduction. It is however not very clear what influenced the present geographical distribution of Nile perch on the continent. Lates has not been found in the lakes of Southern part of Africa because of their general depth and their lower temperature as it prefers medium or higher temperatures.

In Lake Victoria, Nile perch was introduced from Lake Mobutu in 1959, 1962 and 1963 around Jinja by officials of Uganda Fisheries Department and some species from Lake Turkana were introduced by officials of Kenya Fisheries Department at Kisumu in 1963. About eight seedings of perch were introduced at Kisumu point, a number which may have been considered negligible and may not have had any significant biological consequence in Lake Victoria. The main reason for its introduction was to control and convert Haplochromis whose population in the lake was very high into a more desirable and economically viable food crop and extend traditional inshore fishery to offshore waters. Its introduction was also intended to increase the productivity of Lake Victoria as Lates grow to a large size, is extremely palatable and is an excellent sport fish. Apart from those reasons it was believed it would do no harm to Tilapia species because it survives with the same in Lake Turkana.

Since its introduction *Lates* has tremendously increased and spread in Lake Victoria. The first catch was in May, 1960 at Jinja point above Repon Falls. It then began appearing in fishermen's catch in large quantities in Uganda waters. By 1965 it had spread round the Northern shore near Entebbe and Eastwards to Nyanza gulf and Southernwards to Majita Bay. In the Nyanza gulf, the perch began appearing in the fishermen's catch in the late 1960s and early 1970s. In Kenyan part of Lake Victoria the Nile perch is now well established in the gulf with the highest concentration found on a ridge that runs from Homa point to Uyoma point, areas around Ndere islands in Seme and parts of the gulf proper, mainly in the sandy bays. Kenyan Fisheries Department data recording the trends of fishing in Lake Victoria indicate a steady increase of perch fisheries from near zero in 1968 to about 51 metric tons (0.8 %) in 1975 and 27,259 metric tons (about 16 %) in 1981. In 1985 the total Nile perch landing was 50,029 metric tons, 51 % of total landing. In 1987 the landing increased to 86,833 metric tons corresponding to 69 % of the total catches. 1993 estimates from the Kenyan Fisheries department statistical bulletin put the catch of Nile perch to be 99,877 metric tons making it the most abundant and commercially important species in Lake Victoria. The population explosion of this fish in Lake Victoria is attributed to the fact that it is a feeding generalist , and because of its adaptability to global climate change. Other reasons include high fecundity and diversity of habitat as well as the fact that it has a long lifespan. An individual Nile perch may live past 15 years of age.

In general the Nile perch occupies different niches including sandy areas, shallow waters around river inlets and around steep shelving shores. In the Uganda part of Lake Victoria it occurs mostly within 5 meters of water depth, but has been known to extend sparsely to 20 meters depth. In Lake Turkana perch have been caught at a depth of about 50 meters, but it is reported that the species caught at this level was a smaller one known to be endemic to the lake. In the Nyanza gulf high occurrence of perch is limited to 15m depth although it has been caught even at depths of 20 meters. The catches of Nile perch decrease with increasing depth. The fry of Nile perch are normally restricted to inshore areas within the sublittoral weed beds.

Nile perch feeds on a wide variety of different fish species, easily switching to different types and sizes of prey. The major prey of Nile perch when it was introduced in Lake Victoria were detritivorous and planktivorous fishes, but now ecologists believe that it is partly responsible for the disappearance of certain fish species such as *Haplochromine sp.*, *Clarius mosambicus*, *Bagrus docmac* and others in Lakes Victoria and Kyoga. This assertion is supported by the fact that after its introduction, Lake Victoria *Haplochromis*, which formed its main food, decreased from 32% of the total catch in 1977 to less than 1% in 1983 and now they are no longer recorded in commercial catches Likewise since its introduction in Lake Kyoga fish species such as Tilapia species, *Protoptenus aethiopicus*, *Clarius mossambicus*, *Bagrus docmac* and *Haplochromis* have shown progressive decline almost to total depletion during the last two decades of the twentieth century. Nile perch feeds on fish predominantly and in addition feeds on prawns, *Caradina nilotica* and dragon fly,

Odonata species. Lates is also a zooplankton feeder.

Scanty information is available on the reproductive biology of Nile perch. Males mature at a smaller size than the females. Commercial fisheries report peak catches from July to October, but the greatest breeding activity is seen in April just before the long rains. It appears this fish tends not to emigrate far from its territorial grounds for the purpose of spawning. In Lake Turkana the fish that the perch predate on have moved to deeper waters, leaving it on the sandy shallow parts of the lake. The fry which still contain traces of the yolk sac have commonly been found in marginal waters over sandy bottom including aquatic vegetation indicating that these areas are the possible spawning sites. Nile perch show a very high fecundity of about 15 million eggs per spawning. The eggs hatch into many offspring upon which the adults sometimes feed.

Because the Nile perch is a feeding generalist it has the versatility to adapt to a changing ecosystem. Depending on what stage of its life it is in and on the conditions of the lake, it may feed at different trophic levels and within each of these levels on whichever species happens to be abundant. For this reason the logistic equation of the last chapter is not a bad approach to try in modeling the Nile perch. If the logistic equation is an accurate model, we would expect the population of perch to stabilize at a high level. However, we could try to take more subtle effects into account. As we saw with the catfish data, there were large apparent oscillations in the population of catfish, ending finally in a steep decline. Might the perch population eventually oscillate too? If so, it would be a result of factors not accounted for in the exponential or logistic models.

Since the main food resource of the Nile perch at first was the cichlids and since the size of the population of cichlids changed drastically between 1971 and 1980, we might expect that this change had an effect on the Nile perch population. Because it is a feeding generalist, it has been able to survive the drastic decline of the cichlids by eating other species, including its own young. However, if we assume that the perch cannot eat anything but cichlids (which we know is not really the case), then the rate of change of the Nile perch population depends somehow on the number of cichlids available for consumption. In other words, it depends on the amount of available resource. At the same time, the rate of change of the cichlid population depends on the number of Nile perch that are consuming them. It seems as though we will need two differential equations to model the relations between the two populations. Such a set of equations is called a system of differential equations. We will use a standard notation in which N represents the population of a mid-level consumer (here the cichlids) and P represents that of the predator (here the Nile perch).

An equation that describes the cichlid population will have to take into account the rate of change of the population if there were no predator present (a positive rate of change) and the rate of change of the prey population due to predation (a negative rate of change). Very generally for the cichlid population we have:

$$\frac{dN}{dt} = \text{(rate of population increase)} - \text{(rate of population decrease)}$$

If for simplicity we use the Malthusian expression for exponential population increase then where r is a positive constant:

$$\frac{dN}{dt} = rN - p(N, P)$$

The function $p(N, P)$ describes the rate of predation as it varies with changes in the perch and cichlid populations.

A simple version of $p(N, P)$ may assume that the rate of predation is related to the frequency with which a perch and a cichlid come into contact with one another. If we assume that the members of each population move randomly and are evenly distributed, by multiplying the cichlid and perch population densities we can approximate the frequency of their encounters. By multiplying this NP term by a positive constant (b) we can represent the number of encounters that result in prey death. By then subtracting this from the rate of cichlid population increase expected if growth were unlimited (rN) we arrive at the equation:

$$\frac{dN}{dt} = rN - bNP$$

Similarly, the rate of change of the size of the perch population is dependant on its rate of predation because the consumption of cichlids adds to the perch population. The population of perch also depends on the rate of decrease of the population without any prey to eat. Therefore:

$$\frac{dP}{dt} = \text{(rate of increase)} - \text{(rate of decrease)}$$

If there were no cichlids, we could assume that the perch would die at a constant rate because there is no food, and that it therefore would have a death rate that follows a pattern of negative exponential growth $(-gP)$. Like the bNP term in the cichlid equation, we would want to include a term in the differential equation for the perch population that describes the changes in the perch population as the cichlid population changes. Just as the cichlid population lost a certain number of its members, the predator population will gain a number of members proportional to the number of encounters (NP) between the two species. We will use another constant (c) which is the number of predator births that result from each encounter. The resulting differential equation describing the change in perch population is:

$$\frac{dP}{dt} = cNP - gP$$

The set of equations describing the predator-prey relationships, rearranged for more convenient analysis is:

$$\frac{dN}{dt} = rN - bNP$$

$$\frac{dP}{dt} = -gP + cNP$$

The ratio of the constants b and c reflects the relative ease of converting prey into predator, in this case cichlid into perch. A predator like the Nile perch may be commercially more profitable than the smaller fish it eats, but it is higher in the food chain. Because energy stored in its prey is lost in the process of hunting, catching and metabolising prey, it is a less efficient "crop" than an

organism lower in the food chain such as the cichlid. No consumer
is able to convert all of the energy bound up in its prey into energy it
can use. Biologists estimate that at each trophic level 80 percent of
the energy stored in the previous level is lost from the food chain.
Put another way, one kilogram of Nile perch produced means a
corresponding loss of four kilograms of other fish. This estimate of
the rate of energy transfer is incorporated into the equations by the
constants, b and c.

The conditions under which neither the prey population nor the
predator population size is changing (the birth rate equals the death
rate for each population) is called the equilibrium of the system.
To find the equilibrium we set $\frac{dN}{dt}$ and $\frac{dP}{dt}$ both equal to zero:

$$\frac{dN}{dt} = 0 = N(r - bP)$$

$$\frac{dP}{dt} = 0 = P(-g + cN)$$

One equilibrium solution to the cichlid equation is when both
$N = 0$ and $P = 0$. Another equilibrium solution occurs when some
perch and cichlids are present (neither N nor P is zero). In this
case

$$r - bP = 0$$

and

$$-g + cN = 0$$

Solving these equations we find that $P = r/b$ and $N = g/c$.

For each of these equilibrium solutions, the size of each population is constant. In real ecosystems population size does sometimes
fluctuate. What if the size of either or both of the populations
changed a little from its equilibrium value? Would the system go
back to the equilibrium state or would the size of the populations
change drastically? In order to find the answers to these questions
we need to determine whether the equilibrium solutions are stable.

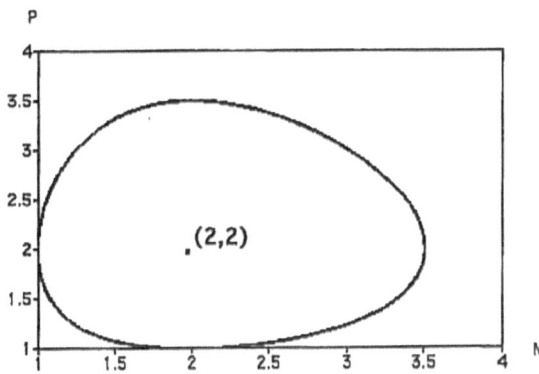

Figure 8.1: Phase portrait for predator-prey equations, Lotka-Volterra model.

The phase plane that we will use to analyze the equilibria of the predator-prey system has N, the size of the cichlid population, as its x-axis variable, and has P, the size of the perch population, as its y-axis variable. This type of phase graph is a combination of phase graphs for the perch and the cichlid equations and is shown in Figure 8.1. We leave out the t axis because the values of N and P at a given time are completely determined by our their starting values. Once again the equilibria of the system appear as points on the phase graph, $(0,0)$ and $(g/c, r/b)$. If we choose the constants $r = 1$, $b = .5$, $c = .5$, and $g = 1$ for our phase graph, the equilibria would be at $(0,0)$ and $(2,2)$ (Figure 8.1).

Now we may use the differential equations of the system to determine the approximate paths of trajectories that start near one of the equilibrium points. A trajectory that begins near the equilibrium point $(0,0)$ could start with $N = 0$ and $P = a$, where a is a small positive quantity. This means that there are no cichlids but there are some Nile perch. We would expect that since the perch have nothing to eat, they would die off. $\frac{dN}{dt}$ would stay at zero and N would remain at zero because cichlids can not be born without parents.

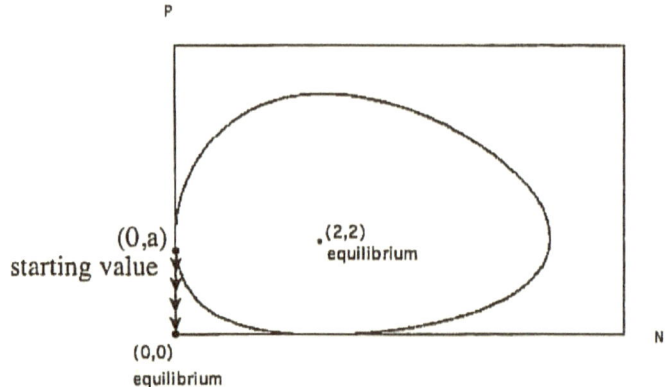

P

(0,a)
starting value

.(2,2)
equilibrium

(0,0)
equilibrium

N

Figure 8.2: Starting at $N = 0, P = a$.

Thus the perch would die off exponentially as a result of our original assumptions. It appears that $(0, 0)$ is stable if we look only at this one starting point. See Figure 8.2. The trajectory starting at $(0, a)$ lies on the P axis.

If instead we had started with the conditions $P = 0$ and $N = a$, we would expect exponential growth of the cichlid population. One of our main assumptions about the system is that the prey population, without any predator, will grow exponentially. Analysis of the equations at $(a, 0)$ shows this to be the case. $\frac{dN}{dt} = rN$. It appears that $(0, 0)$ is unstable if we look only at this one starting point, as in Figure 8.3.

If both the perch and the cichlid populations start out with small positive values ($N < g/c$ and $P < r/b$), $\frac{dN}{dt}$ would be positive and $\frac{dP}{dt}$ would be negative. The perch would not have enough to eat and so it would die off. At the same time, there would not be enough perch to prey on all of the cichlids. The cichlid population would grow, but at a slower rate than it would without the presence of any Nile perch.

The fact that most solutions tend away from $(0, 0)$ but at least one tends towards it shows us that this unstable equilibrium is something mathematicians call a saddle point.

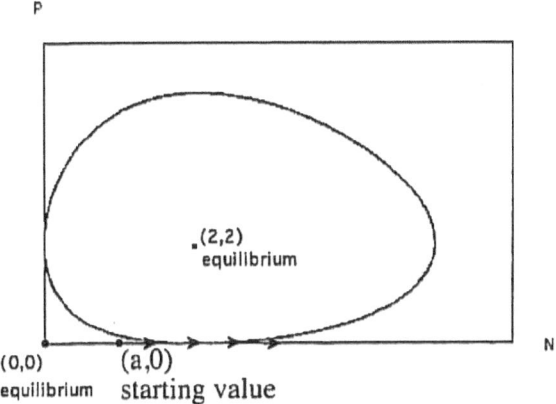

Figure 8.3: Starting at $P = 0, N = a$.

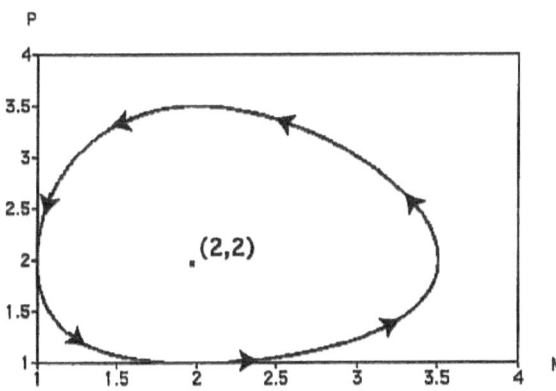

Figure 8.4: Typical trajectory for solution to Lotka-Volterra: Perch versus cichlid.

If we test small divergences from $(g/c, r/b)$ we get the trajectories shown in Figure 8.4. This equilibrium point seems to be a center point because the trajectories appear to form closed loops around it. The computer program used to generate this graph starts with one value of N and one value of P and then plots the approximate path of the trajectory by using $\frac{dN}{dt}$ and $\frac{dP}{dt}$ to estimate the changes in N and P that give us the next point. What direction are the trajectories traced over time? We can use the differential equations to find the direction of change of N and of P. If we start with $P < r/b$ and $N < g/c$, $\frac{dN}{dt}$ is positive, and N increases. However, $\frac{dP}{dt}$ is negative and P is decreases. If $P < r/b$ but $N > g/c$, both N and P increase because both $\frac{dN}{dt}$ and $\frac{dP}{dt}$ are positive. If $N > g/c$ and $P > r/b$, N increases and P decreases. Finally, if $N < g/c$ and $P > r/b$, both N and P decrease. These qualitative changes show us that the trajectories are traced in a counterclockwise direction around the equilibrium point. The more sophisticated approach the computer uses indicate that the solutions oscillate around the equilibrium but they never reach it.

The cichlid and perch population sizes will change between values above and below those of their equilibrium. Each will pass through its equilibrium value, but never both at the same time, so the system will never be at equilibrium. We can see these oscillations if we plot the Tanzania catch data for Nile perch and cichlids against time, as shown in Figure 8.5. A phase graph of the data is shown in Figure 8.6.

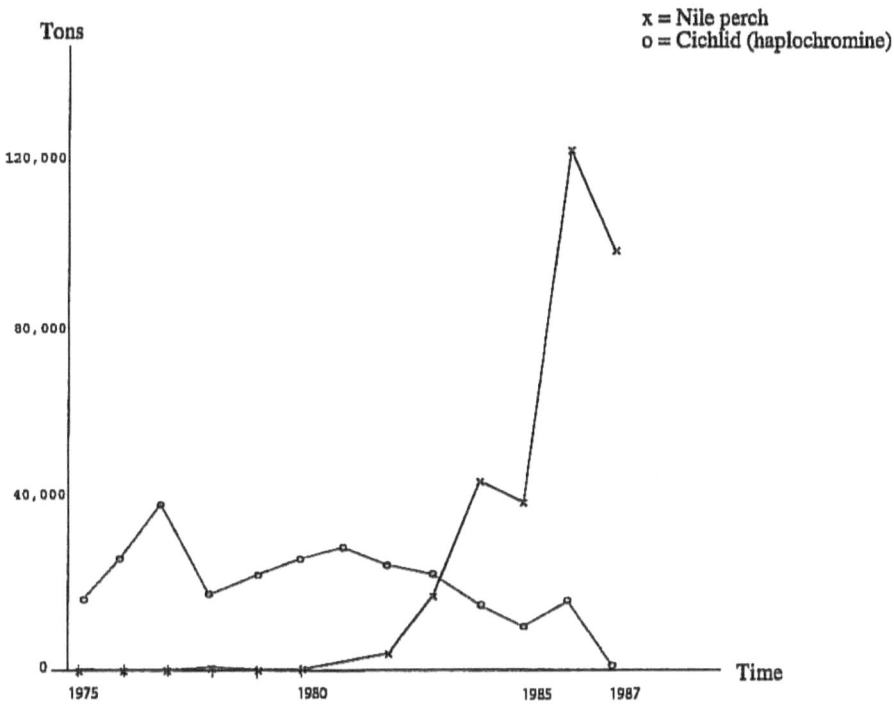

Figure 8.5: Perch and cichlid versus time.

Figure 8.5 and 8.6 illustrate the utility of the phase portrait. If we only looked at graphs of the perch with cichlid populations over time, we might conclude that they oscillate more or less as predicted by the Lotka-Volterra (predator prey) equations we have been studying. Looking at Figure 8.6, however, tells us that our model is very far from what the data are doing. We already knew the perch ate more than one prey fish, but if we didn't know that then comparing the phase graph of our model against the data would be a big clue that we had overlooked some factors.

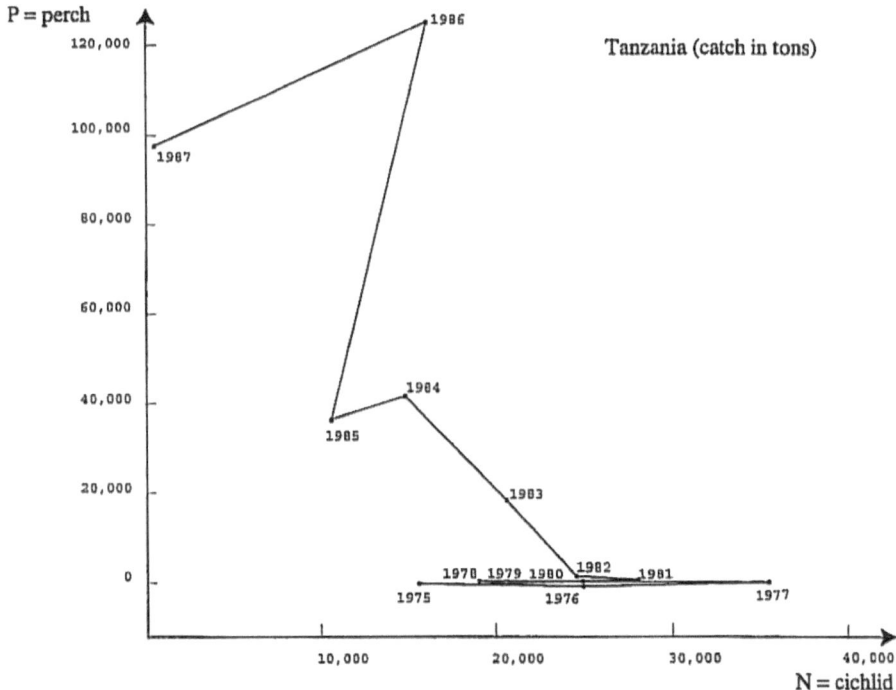

Figure 8.6: Perch versus cichlid.

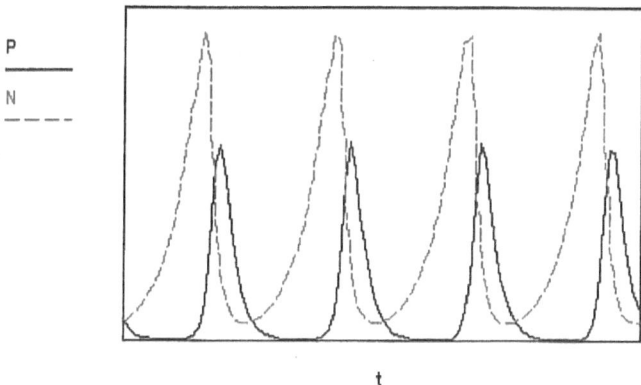

Figure 8.7: Typical solutions to Lotka-Volterra plotted against time.

Figure 8.7 shows solutions to the predator - prey plotted over time. As you see, it is harder to see from this kind of graph that something is drastically wrong with our model. Although the model captures the ups and downs of fish populations, it does so in a periodic way that is not particularly close to what the data show. As with the logistic equation, we have been able to model a general qualitative feature of the system but with poor ability to predict specifics. For fisheries catch data, this failure is not surprising, and this is probably the right place to talk generally about fish models.

During its life cycle a typical fish may play the role of prey, predator, or both. Many species of fish lay eggs, which are prey for a large variety of organisms that may themselves be prey for hatchlings, fingerlings. or older versions of the same fish. Some fish, such as the Nile perch, are such feeding generalists that they routinely consume many of their own offspring. So a more complete model for the Nile perch would include boxes for different sizes of perch, with species interactions specific to those sizes. Furthermore, one predator left conspicuously out of the model in this chapter is human. Catch data, in particular, is not just a benign measure of the fish population but the result of large scale predation on perch by human beings. Our predation is related to the size of the fish population, but is also controlled by a variety of regulations concerning seasonal control, daily catch limits, mesh size of nets and other factors. Further, in Lake Victoria both the perch and their prey are fished, but in different locations, with different methods. The human predator has different behavior with respect to different species. A more useful model would take some of these factors into account. What follows in this chapter is a brief description of the fishing industry in Lake Victoria at the end of the twentieth century.

The effect of the introduction of Nile perch on the ecology of the lake was matched by the effect of improved fishing techniques in the region. Together these changes had enormous effect on the local economy. Larger catches, sophisticated equipment, processing plants and a commitment to export drastically changed the way everyone in the region made their living.

The fishing gear drastically changed from simple traps to sophisticated indiscriminate methods that included the use of trawl nets, complicated gill nets, long hook lines with better ways of lowering fish and wide beach seining facilities. The fish that were fully in demand by this time included Tilapia species, catfish, (*Carius mossambicus*) and lungfish because of its tasty firm meat. However with the introduction of dairy, poultry and fish farming, *Rastreoneobola argentia* (*omena*, sardine) which is massively used as animal feed soon became the focus for commercial beach seining. The seine for *omena* is intended to catch small sized fish and hence could not discriminate the small *haplochromis* and juveniles of other fish species. It is likely that the continual use of this type of gear could have contributed to the disappearance of most of the fish species because it indiscriminately catches the fingerlings which are crucial for the procreation of species. In the early 1970's as much as 35 percent of the commercial catch was immature. Researchers concluded that the larger *haplochromis* had almost been eliminated by overfishing. The *omena* industry thrived, thereby encouraging processing plants. The processing of this fish for animal feed involves the drying of the fish for several days followed by grinding and packaging. *Omena* is also a rich source of proteins and has been used in hospitals for protein replenishment of malnourished children. Since it was being caught in tons it became popular with the public as the cheapest source of fish protein. Rich individuals invested in fisheries and sooner or later the catch had to dwindle with or without the introduction of other factors into the ecosystem.

Although pollution and introduction of Nile perch might be blamed for the disappearance of several species, the rate of over-fishing and indiscriminate fishing without corresponding stocking of the lake has quite possibly contributed more to the disappearance of several lake Victoria species. With the systematic decrease in the population of other fish species the population of Nile perch dramatically increased. The rapid increase in the population of this fish has been attributed to its high fecundity and the availability of varieties of food within the lake. The population explosion of this fish brought in an alternative for fish processing industries whose survival were now threatened by low catch of *omena*.

Several Nile perch processing plants have mushroomed around lake Victoria with high concentration on the Kenyan side. The processing of Nile perch starts immediately when it is caught. Due to high temperatures the boats must be equipped with cooling facilities to reduce spoilage during long fishing episodes. The fishermen who lack these facilities are heavily exploited by middlemen who provide cooling facilities in exchange for low prices. Several cooperative societies established with the intention of saving the fishermen from this kind of exploitation have failed. The fishermen therefore are at the mercy of the middlemen and the large processing plant owners. The roads to the Nile perch collection points are poor. The refrigerated trucks get stuck during the rainy season, leading to spoilage of fish before they reach the processing centers. There is no electricity supply, treated water and telephone connection to these centers. It is therefore unfortunate to realize that the exploitation of the Nile perch by these processing plants does not necessarily improve the economic status of the local community.

Chilled whole Nile perch are landed on the beach where they are weighed and transported to the processing plants which are located in the major cities, especially Nairobi and Kisumu in Kenya. Kisumu has the highest concentration of fish processing plants since it is located on the shores of Lake Victoria. The processing of Nile perch in the factory starts by filleting, and deskinning . The fillets are taken to the trimming section where they are into 2kg size for easy packaging. They are then packaged and stored into huge coldrooms awaiting exportation. The fillets processed this way are too expensive for the local market and are exported to Israel and the Middle East, the United Kingdom, and other western countries. The factory owners at one point paid approximately $0.1 per kg for unprocessed Nile perch fillet but sold it at about $5 or more per kg, thereby making a large profit. Processing factories mainly utilize the flesh as fillets for source of proteins, although no part of the Nile perch is wasted. The skin is used for making leather, the air sack for making surgical threads, the fats from the abdominal cavity for cooking oil and the carcass (mgongo wazi) sold back to the small scale business women who fry it and sell it locally. The local community which participates in the conservation of the lake is finally given the carcass simply because it cannot afford the fillet.

The processing plants are complex facilities employing large numbers of workers. Most often they are unskilled laborers working for a day's wage. Most of the workers are hired on temporary basis (as of 2000, when this description was written), paid low wages, are not entitled to any benefits and do not belong to any union which can defend their rights. Due to an unskilled inconsistent labor-force and undefined divisions of the compartments of the process-ing stages, there have been several reports of cross contamination of the finished products with such deadly bacteria such as E. coli and salmonella. A lot of consignments have been rejected and destroyed in the countries of destination. Because of this, the UNESCO Food and Agriculture Organization (FAO) installed a set of hygienic con-ditions that must be met by the processing plants to be allowed to export. These conditions included employment of regular trained in-dividuals occasionally checked by medical personnel, reorganization of the plant so that the offloading zone for whole fishes is separate from the finished products, presence of treated running water and sanitized floor and other facilities. The factory is supposed to be inspected by Fishery officers at least twice a month. Plants that meet these requirements are allowed to use specific stamps on their finished products to allow them to export to European countries.

The export of Nile perch has recently become one of the major foreign exchange earners for Kenya. However with the high rate of exploitation there is a likelihood that the Nile perch business might soon be faced with eminent extinction. There are several ways by which this industry can be maintained. One suggestion would to create certain fish breeding zones within the lake where absolutely no fishing is conducted. The other alternative would be to continually restock the lake with not only Nile perch but also with other fish species which are rapidly disappearing. Since the Nile perch business generates a lot of income, encouraging aquaculture is also a possibility. However this will require large areas or large ponds, since this fish grows into very large sizes.

For more information on Nile Perch, and modeling their population, see Mugisha & Ddumba (2007),which includes the constants used for Nile Perch models, Ogutu-Ohwayo (1990), and Ogutu-Ohwayo (2004) for Nile Perch models with an explanation of political ramifications, and data in the latter article. For a broader base of understanding of Nile Perch and other introduced fish in Lake Victoria, without modeling, Njiru *et al* (2005) provides a reference. For a better understanding of fish population changes in lakes other than Lake Victoria, and with other fish, DeAngelis *et al* (1997) and Wu & Culver (1994) give models with relevant parameters which can bring insight to modeling Nile Perch, or understanding the mathematics of fish in general.

For your consideration

Question 1:

The relationship between human and Nile perch is also one of predator to prey. Model this relationship using a pair of equations describing change in each of the two populations. What are your assumptions about fish and human?

Question 2:

How does the nature of the model change depending on whether the exponential or logistic equation is the basis of the model? Which population (predator or prey) does it make sense to cap with a logistic term? Use a computer to draw phase portraits of your system and investigate the long term effects of these two assumptions.

Question 3:

How would you modify your model to include the lower level prey (haplochromis, or cichlid)? What if a major portion of the adult perch's diet is juvenile perch? What if cichlids and other small fish also eat perch eggs? There is a web of relationships here that can be sorted out with assistance from the literature.

Question 4:

One of the most challenging aspects of ecological models is determining the constants of growth, predation, energy transfer up the trophic level, etcetera. Pick a variation on the model in this chapter and go to the literature to find these for a specific situation.

Question 5:

How would you adjust your human-perch model to account for the restocking of the lake with perch, as is suggested by various people? How would you account for the designation of no-fishing areas, as is also suggested? Be sure to state your assumptions about fish behaviour carefully and take them into account in your equation.

Chapter 9

How to think about a model and go fishing

The poor match between catch data and the predictions of the predator-prey model of the last chapter is a result of multiple factors. In particular there are three difficulties worth thinking about with regard to that model.

First of all we have to question the reliability of catch data as a reflection of the fish populations. As the chapter pointed out, during the years that data was collected, fishing underwent major changes in the region. A broader range of fish were caught and more resources were put into catching the larger perch. Fish populations may not have been quite as erratic as the data suggest and some of the randomness may be the result of changes in fishing rather than fish.

Second, the environment and therefore the species living in it are subject to random effects of weather, human activity and other forces. Only in controlled laboratory conditions do we tend to see models conforming closely to actual data. The real world is messy, with random shocks to an ecosystem being the norm rather than the exception. Of course it is possible to model random changes. One could build a program that randomly enlarges or diminishes a population by a small amount at every time step of the numerical algorithm. Done correctly, this would result in an output that looked a bit more like real data in general but not more like any particular data set.

Third, we may have made a model that failed to incorporate some important assumptions about the situation. This is certainly true of the model in the last chapter. The cichlid is not one single species, but many. It competes for resources with other small fish, none of which are in the model. The Nile perch is a feeding generalist and eats those other fish, so it is not solely reliant on cichlid, unlike what our model would have us assume. The Nile perch even eats its own juveniles, functioning at times as both predator and prey. So there are lots of aspects of the real situation that are not accounted for in the model.

The utility of a model does not always lie in its ability to match a data set, especially one heavily influenced by more or less random events. We could always try to fit a function to a set of data points and there are many ways to do so. If we were to fit a polynomial to the data set in the last chapter, it might fit perfectly. In the long run, though, it would behave like a polynomial, going up infinitely far or else becoming negative and dropping. Alternatively, we could fit a sum of sines and cosines of different amplitudes and frequencies. If we used enough of them, we could fit the data exactly. In the long run, our function would oscillate endlessly, because of the choice of functions we used to build it. Then we would have to decide whether we believe the polynomial answer that (let's say) rises indefinitely or whether we believe the oscillating answer. Our beliefs are, of course, not a reliable method of predicting the future. If they were, we wouldn't bother to make a model at all.

The point of modeling with differential equations is that we build our assumptions about how nature works into the equations themselves as hypotheses. The solutions then tell us which macroscopic features will follow from those hypotheses. In the last chapter one hypothesis was that growth in predator population depended on quantity of prey available, according to a rule that we prescribed. The first part of this hypothesis is not a huge assumption but the actual growth rule we used could certainly be adjusted. We also assumed that the prey was diminished in a way that depended on the number of predators. Also perfectly reasonable, except that the actual rule we prescribed might be improved. Finally we made assumptions about what these two populations would do in the absence of the other one. If there is no prey, the predator follows the rule:

$$P' = -gP$$

The predator would thus die out exponentially. If there is no predator, then the prey follows the rule:

$$N' = rN$$

The prey would thus grow exponentially. Together the full set of equations was:

$$N' = rN - bPN$$
$$P' = -gP + cNP$$

So these are our hypotheses. The long term behavior predicted by the model has two main features. First of all, there is an equilibrium solution. Second, all of the other solutions are periodic, oscillating forever. In fact, oscillations of all sizes are possible, with populations of either predator or prey getting arbitrarily large.

The predator-prey equations cleanly demonstrate the difference between modeling with differential equations versus just fitting functions to data. In the differential equation model we make assumptions about how nature works now, and see what the model predicts in the future. To fit a curve to data we don't need to make any assumptions, but if we want to know what will happen in the future we have to assume we know what the future will be in advance of choosing which functions to use to approximate our data. The first kind of modeling is more powerful because our assumptions are about what is happening in nature now, therefore they can be tested, observed, and debated more usefully than our guesses about the future.

Of all the hypotheses we built into these equations, the one easiest to argue about is that the prey population is capable of limitless growth. Nowhere in nature do we observe a species to have this property. Other than economists discussing the stock market, nobody ever believes limitless growth of anything to be a reasonable prediction for the future. But in the lab, we do sometimes see growth following a logistic curve. So we could argue pretty convincingly that, even though it won't be perfect, a model that results in logistic growth for prey in the absence of predator would be a better hypothetical situation than what we have used so far. To do this we need only change one term:

$$N' = rN(1 - N) - bPN$$

$$P' = -gP + cNP$$

Now if there is no predator, N follows the rule:

$$N' = rN(1 - N)$$

This is the logistic equation from chapter 4. The growth of N is limited by the number 1, the carrying capacity of its habitat. Of course we could have chosen a different carrying capacity, but for the purposes of comparing models this will do for now. Think of the units of N as being "percent of carrying capacity". Here is one solution typical of this model, first displayed as a time series and then as a phase portrait in Figures 9.1 and 9.2.

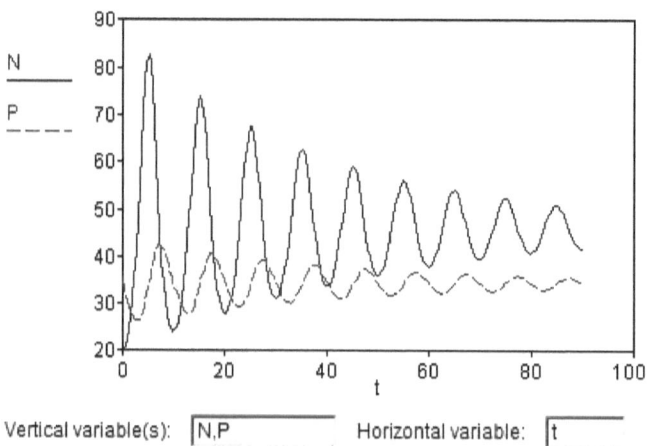

Figure 9.1: Time series.

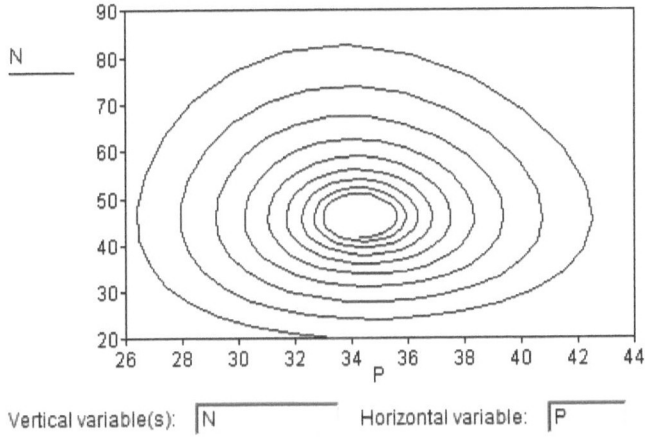

Figure 9.2: Phase portrait.

An important qualitative difference has resulted from changing just one assumption. Whereas our original predator/prey equations had solutions that oscillated forever with no reduction in amplitude, these solutions show damped oscillation, tending eventually to equilibrium. Comparing these two models supports the assertion that limits to growth are a factor that causes populations in a predator/prey relationship to approach equilibrium.

The tendency of populations to reach an equilibrium value and stay there is the subject of much discussion among ecologists and modelers. Often we assume that an ecosystem was in equilibrium until we humans came along and disturbed it. There usually isn't any data to support such an assumption; just the general belief that, left alone, a system would behave thus. Reasoning completely backwards (which we should never do) we would prefer the damped predator/prey equations because they produce an answer we would like to believe. Fortunately we don't have to reason backwards because the assumption of limits to growth is far easier to justify and, in this case, it produces solutions tending to a limiting equilibrium, a property that we desire.

Returning to the discussion of random events at the start of this chapter, we might investigate how the two systems we have been studying would respond to such a perturbation. Figure 9.3 is the phase portrait for the original undamped predator/prey equation.

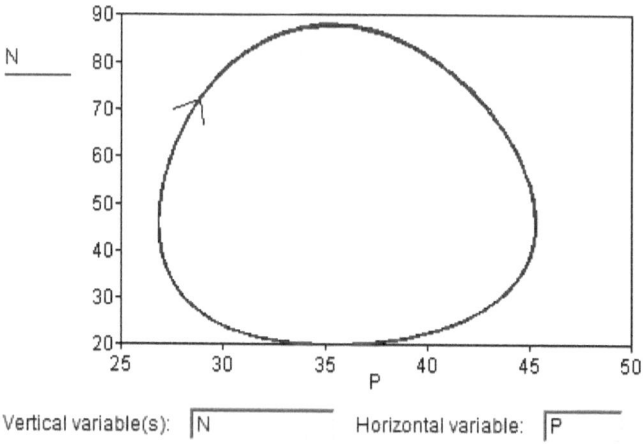

Figure 9.3: Undamped phase portrait.

If an event, (such as fishing), were suddenly to lower the population of predator, one of two qualitatively different things would happen. Figure 9.4 shows two possible points where a lowering of predator would have different effects.

If the predator were removed at point A, the solution moves to a trajectory outside its original one. It moves farther from the equilibrium value and it oscillates with greater amplitude indefinitely. If the predator is removed at point B, the solution moves inside its original trajectory and oscillations are reduced in amplitude. Note that at A and B the prey population is exactly the same. In this model, a random disturbance is propagated throughout time. Figure 9.5 shows the time series for these two situations.

Notice that the perturbation is propagated throughout time. That is, a single removal of a small amount of one population creates a change in amplitude of the oscillation that is propagated forever. By contrast, the damped predator prey equation the phase portrait looks like Figure 9.2, with trajectories that spiral in to a fixed point. Once again, we could reduce the predator at one of two locations. But because all solutions spiral in toward the equilibrium solution, eventually traces of the random event will disappear, as shown in the time series plots, as shown in Figure 9.6.

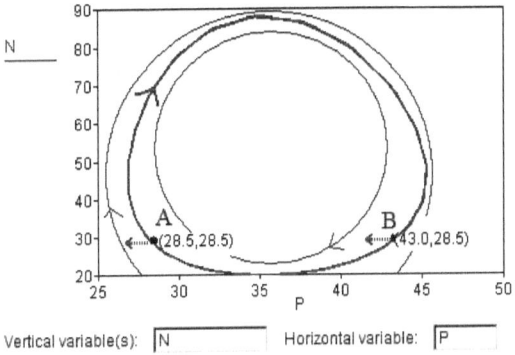

Figure 9.4: Undamped phase portrait showing the effect of perturbation.

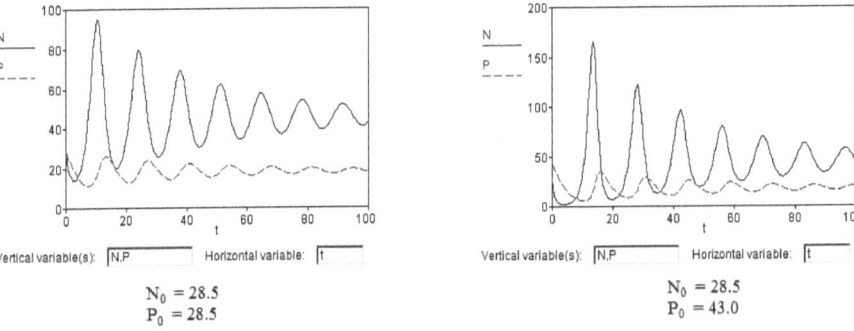

Figure 9.5: Damped time series showing the effect of perturbation.

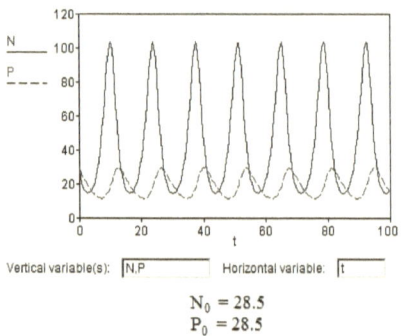
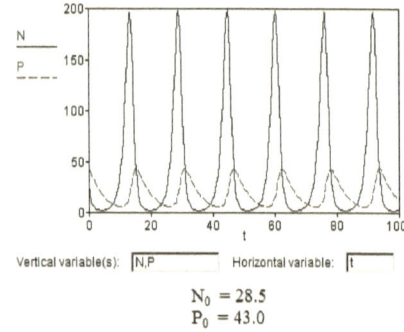

$$N_0 = 28.5$$
$$P_0 = 28.5$$

$$N_0 = 28.5$$
$$P_0 = 43.0$$

Figure 9.6: Undamped time series showing the effect of perturbation.

For these models, the introduction of an assumption of limited growth is actually a mitigating factor against random perturbations. In both cases an ill-timed reduction of predator population can result in a rebound effect where the predator population become larger than it would otherwise have. But in one case the effect returns in every cycle, whereas in the other case it dies out. a population will have different effects depending on where in the population cycle it is done. Sometimes the point of harvesting is to maintain an equilibrium (for example moose license policy in New Hampshire), sometimes it is to maximize yield over time (in the fishing industry for example), and sometimes the point of harvesting is extinction of the species in an area, at least for a while (such as Japanese beetle in my garden or *Anopheles* mosquito in malaria infested regions). If the predator-prey models in this chapter are good qualitative descriptions of the way the organism grows then they can be used to design optimal strategies for harvesting that organism, even in the face of random disturbances of the system.

For your consideration

All of the questions at the end of Chapter 8 remain pertinent here. It is worth revisiting them using a damped version of the predator prey model to see what qualitative or quantitative effects are changed by the assumption of limited growth of prey.

Chapter 10

Predator Satiation

What does it mean to eat well? By now we should know what nature has always known. Eating well means eating as much as you need, not as much as you can. For humans this is an important distinction, but for most species the two amounts are often the same. It is efficient to have a stomach capacity that accurately reflects the calories needed for survival and reproduction. A stomach that holds a lot more might reduce the organism's capacity to do other things. Most creatures stop eating when full (and so should we).

In the last chapter we looked at the growth term for prey in the predator/prey equations and replaced it with a logistic term that limits the growth of prey in the absence of predator. We ended up with these equations:

$$N' = rN(1 - N) - bPN$$

$$P' = -gP + cNP$$

Now let us look at the terms that respond to the rate of predation, bPN and cNP. For a fixed amount of prey, N, this term is just proportional to the number of predators, P. This seems reasonable because doubling the number of predators would probably double the chances that a prey organism would be caught.

For a fixed amount of predators, P, the same term is just proportional to the number of prey, N. It says that doubling the amount of prey will result in twice as much prey being eaten. If there are a lot of predators relative to the amount of prey, this would certainly make sense. The predator will eat as much as it can catch. But if there are only a few predators and a huge supply of prey, it makes less sense. The prey would eventually be constrained by its own lack of hunger or the amount of time in the day or some other factor and would be unable to continually double the amount it eats indefinitely. The predator would eat at most as much as it needs. So at low prey levels the rate of predation might rise proportionally but at higher prey levels it would become constant.

This general phenomenon is known as "functional response" and it comes up in a lot of other contexts. For example when your immune system responds to an infection it produces white blood cells in response to the quantity of infection present, but there is a maximum rate at which it can do so. A single heart cell admits calcium ions through molecular gates. At small concentrations of calcium it does so in proportion to the concentration, but at higher concentrations the rate of entry is limited by the number of gates through which the ion may pass. A similar functional response holds for calcium transport.

To take the predator/prey situation as an example, we might wish to replace these equations:

$$N' = rN(1 - N) - bPN$$
$$P' = -gP + cNP$$

With new equations of a general form

$$N' = rN(1 - N) - bP \times f(N)$$
$$P' = -gP + cP \times f(N)$$

Where $f(N)$ is some function of the prey population that starts out looking like "N" but after awhile approaches a constant, say "1". So at low populations of prey the term $cP \times f(N)$ is approximately cPN but at high prey populations it is approximately cP. So we need a function $f(N)$ that looks like N to when N is small but like 1 when N is large. Put in the language of mathematics, we would say that the limit of $f(N)$ as N goes to infinity is 1, and also the limit of $f(N)/N$ as N goes to zero is 1. Another way of saying the second limit is that "$f(N)$ is on the order of N as N goes to zero" or even "$f(N) = O(N)$ at zero". This means that not only does $f(N)$ approach zero near $N = 0$ but it does so with slope 1, like N.

Figure 10.1 is a picture of what f ought to look like.

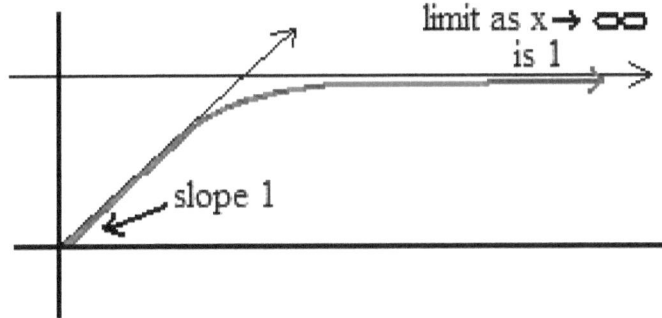

Figure 10.1: A desirable functional response for a predator.

There are a lot of functions that can look like this. Here is the simplest: $f(N) = N/(1 + N)$, shown in Figure 10.2.

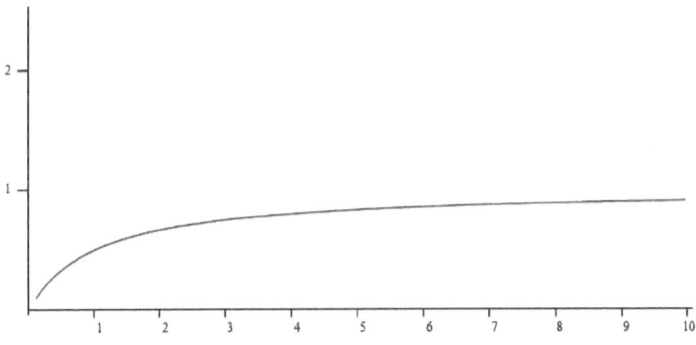

Figure 10.2: Graph of f(N)=N/(1+N).

Sometimes our assumptions of predator behavior are more nuanced. Perhaps we believe that if N is very low the predator won't bother to hunt it. This would hold especially if there are alternative prey sources, as in the case of the Nile perch. Then we want a function whose slope is zero near zero but rises to a limit of 1, as in Figure 10.3.

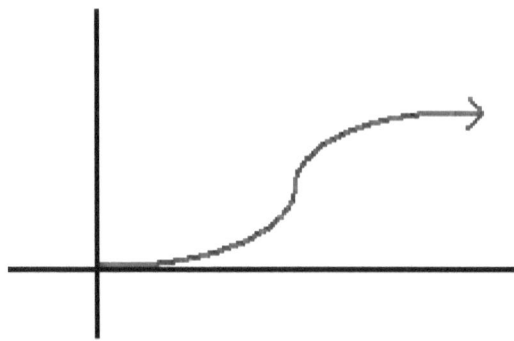

Figure 10.3: Another possible function response for a predator with alternative prey.

Functions of the form $f(N) = N^k/(1 + N^k)$ all do this job well, The exponent, which must be greater than 1, controls how quickly the function rises to 1 (see Figure 10.4). When the predator is eating as much as it possibly can, the constant "b" becomes the amount of prey per predator per time unit that perch with unlimited access to prey are able to eat. For Nile perch, one study observes perch to eat once a day, with a full stomach containing 1 to 8 percent of the weight of the perch in prey fish. A modeler would use this study to figure out the constant "b". It can be tricky to estimate this in terms of absolute units such as carrying capacity. The food web supports fewer top predators (in terms of biomass) than prey animals, so one might invoke rules of thumb to get such estimates. The constant "c" then represents the proportion of weight of prey consumed that is actually converted into predator biomass in the next generation.

Modelers use a variety of functions to capture these different assumptions about predator behavior or other rate-limited processes. In some cases, such as calcium transport across a cell membrane, the process can be measured directly to give both a functional form and accurate constants. In other cases, such as ecology, the form of the function just embodies assumptions about the behavior of an

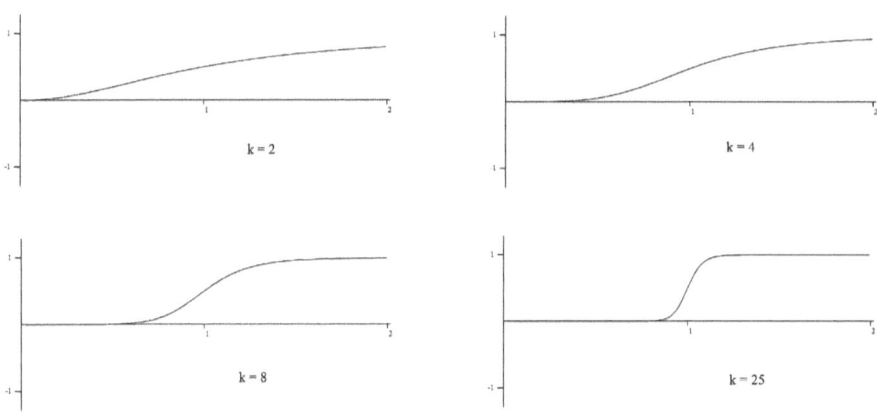

Figure 10.4: $f(N) = N^k/(1 + N^k)$.

organism and it may not be possible to get constants directly from data.

In the literature these functions are sometimes called "Holling functions" or "Michaelis-Menton functions", depending on the field but usually in ecology or medical literature. Mathematicians do not usually give special names to particular simple functions. They would call f a "rational function" when it is the quotient of two polynomials (when k is a positive integer).

Ecology models that incorporate a functional response of this sort for predators are sometimes use to model a phenomenon known as "predator satiation". This situation occurs when the prey organism procreates so abundantly that its predators cannot possibly consume a proportional number of offspring. It is usually cited as an adaptive mechanism to avoid predation. To see the adaptive value of this strategy we could compare two models:

Model 1: logistic growth of prey and strictly proportional predation

$$N' = rN(1 - N) - bPN$$

$$P' = -gP + cNP$$

Model 2: logistic growth of prey and predation following a Holling functional response:

$$N' = rN(1 - N) - bP(N/(1 + N))$$
$$P' = -gP + cP(N/(1 + N))$$

Figure 10.5 shows the outcomes for these two models.

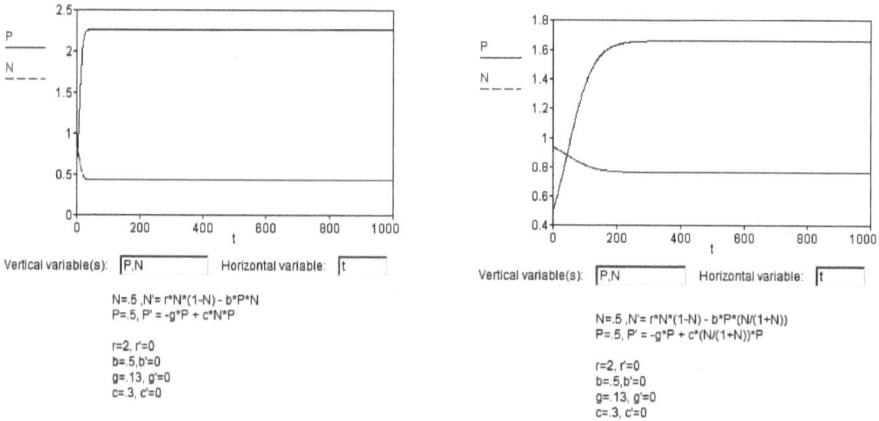

Figure 10.5: Model 1 on left, model 2 on right.

In this case the models are compared in order to justify the statement that predator satiation does indeed have adaptive value to the prey organism. A complete study would vary the growth rate of the prey (the constant r) to see how it affects long term behavior.

A model of this sort might be used to justify a statement about evolution itself, inferring very long term consequences by the limiting behavior of a system which, although a "long term" consequence of the system, is still a short time frame compared to that of evolution. Notice, for example, that switching from model 1 to model 2 increases the equilibrium value for the prey and decreases the equilibrium value of the predator (Figure 10.5).

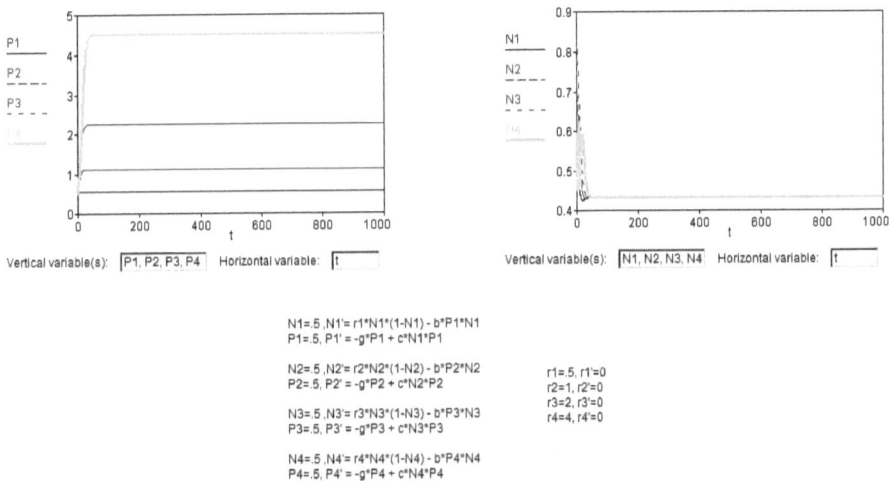

Figure 10.6: Model 1 with varying choice of r. Predator on left, prey on right.

The constant r does not affect the equilibrium value of prey in either model (Figures 10.6 and 10.7). The comparison of what happens to equilibrium values of predator is displayed for both models but not very elegantly. A better visual display would be a plot of r versus equilibrium values for each of the two models. The effect of the other constants on these models might also be interesting. Finally, a good modeler would solve explicitly for equilibrium values in both models if possible, which would give an analytic expression for the dependence of equilibrium states on all parameters.

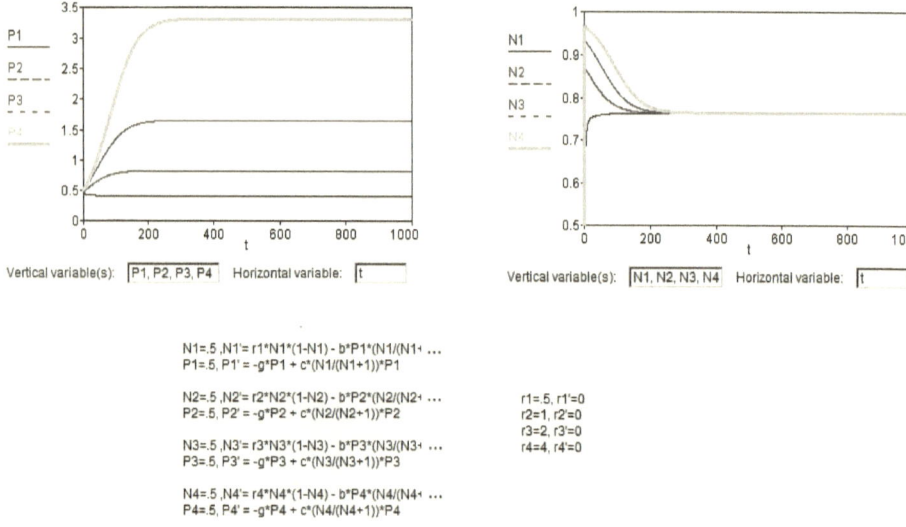

Figure 10.7: Model 2 with varying choice of r. Predator on left, prey on right.

For your consideration

Exercise 1:

Use data about feeding and reproductive habits for perch and general knowledge about energy transfer in the food web to estimate the constants in the models in this chapter.

Exercise 2:

Some species reproduce seasonally. In this case "r" is not really constant, but oscillating. One could model "r" as an oscillating function of time. Do this.

Exercise 3:

What is a better strategy for prey? A continued high growth rate or an oscillating growth rate? What output of your model is the best measure of "better" in this case?

Exercise 4:

The Nile perch produces offspring as eggs which mature into juveniles and later into breeding adults. The typical life span is around 8 years, with 3 years spent in the juvenile non-reproductive stage. Juveniles eat prawns, mollusks and insects whereas the adults eat cichlids and other small fish. The adults eat the juveniles if they can find them. (According to some sources up to 60 percent of stomach contents of an adult perch have been found to be juveniles. Perch can eat 1-8 percent of their own weight per day.) Maturation of juveniles depends on the number of juveniles and the carrying capacity of the larger system for juvenile perch. It is reasonable to assume the predation of adults on juveniles obeys a Holling functional response. Build a model for this situation where the only prey available to the adult perch is its own juveniles.

Exercise 5:

For the model in problem 3, what parameters give long term continuation of all populations? That is, what (if any) growth and maturation rates will allow the Nile perch to survive while living entirely off its own young?

Chapter 11

Mbuta

Africa is the birthplace of humankind, to the best of our knowledge. The original Garden of Eden, she is the land from which all human migration began. With our current scientific understanding, we believe her to be the source of the human race.

Let us turn this observation on its head. From the point of view of the myriad other organisms inhabiting the African continent, Africa is the place humans have inhabited the longest. Her ecosystems are those first to include humans. Her people and her ecosystems have co-evolved for longer than any others on this planet. From Africa came the original human diaspora.

In the area where a new species evolves, the ecosystem evolves along with it. The place where the Nile Perch evolved likely included a complex system of predators, parasites and diseases that co-evolved with it. But when the perch was introduced to foreign waters, such as Lake Victoria, it was in a special position. It may have brought its own diseases and parasites with it, but it left its predators behind. The niche it occupies in a new location used to be taken up by one or more indigeous species that had no competitors until the perch arrived. In particular, the Bagrus Catfish found itself in competition with the Nile Perch for smaller fish.

We would expect a similar but more dramatic scenario when humans arrive in a new location. Some have argued that, in their place of origin, early humans did indeed have natural predators, although these arguments may be rather speculative at this point.

In any case, when humans moved out of that first ecosystem and into a new one, their predators probably did not move with them. This situation gave them a competitive advantage with other species. Modern human beings are omnivorous. Although we don't always know what their early eating habits might have been, it is likely that they were strong competitors with indigenous species for a variety of food sources. Some have pointed out that periods of mass extinction in certain locations loosely coincide with the arrival of human beings at those locations. Although this observation proves nothing, it is consistent with our understanding of what happens when species are introduced to a new ecosystem.

Our understanding of this phenomenon rests squarely on mathematical models, as controlled experiments on the scale of whole ecosystems are impossible. Why might the introduction of a new species cause the extinction of at least some of its competitors? Does it always work that way? If not, what circumstances must be in place for both species to co-exist? Our model is built from a simple premise. In isolation, without natural predators, each of two competing species will grow in population according to the logistic equation we looked at earlier. Let us use P for perch and C for catfish. Separately, these equations look like this:

$$\frac{dP}{dt} = aP(1 - P)$$

$$\frac{dC}{dt} = bC(1 - C)$$

When we put those two species in the same lake, each of their equations gets a new term added to reflect the disadvantage it has in the presence of the other population. If the perch population is constant, an increase in catfish puts a negative strain on it. If the catfish population is constant, then an increase in perch also puts an extra strain on the perch population, in proportion to the number of catfish present. So, the term we add to both of these equations will be a product, PC, as in the Lotka Volterra model. You might say that P represents the chance a perch will be hunting in a certain area and C represents the chance a catfish is hunting in a certain area. If their behaviors are independent of each other, then PC is the chance both of them are hunting in the same region. That is the situation that results in competition and a disadvantage for both species The resulting competition equations look like this:

$$\frac{dP}{dt} = aP(1 - P) - mPC$$

$$\frac{dC}{dt} = bC(1 - C) - nPC$$

Of course, the disadvantage may not be fair, so in both cases there will be a constant in front of this term, and the two constants may not be the same. Competition between human beings and other top predators provides an excellent example of a situation where the disadvantage is likely to be unequal. When humans enter a new region as hunters they might compete with lynx, let us say, in the taking of smaller animals for food. Now, the lynx is completely dependent on a supply of these small animals for its diet, whereas humans are omnivores. If rabbits are in short supply, the people may rely more on fish. In extremely dire times they may rely completely on vegetables for a fairly long duration. In modern examples, humans as farmers may maintain a supply of their own protected species that the lynx cannot touch. In any case, an increase in humans will have a far larger impact on the lynx than an increase in lynx will have on humans. For the equations describing a competitive interaction, the constants will be very different for the two populations.

As in the Lotka Volterra model, we will analyze these equations using a two dimensional phase portrait. For these equations, something very different happens. You should remember that, in the Lotka Volterra model, when each of the derivatives is set to zero the necessary consequence was that either one of the populations was zero or else both populations were at the fixed point or equilibrium . Now we are going to look more closely at the situation where only one of the derivatives is set to zero. The locations where dP/dt is zero in the equation above look like this:

$$C = \frac{-a}{m}P + \frac{a}{m}$$

In addition to the line where P is extinct, there is another line in the phase portrait, called an isocline, pictured in Figure 11.1.

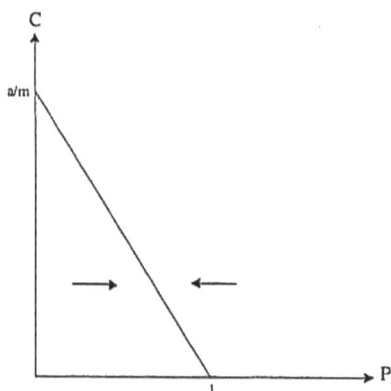

Figure 11.1: $dP/dt = 0$ on the slanted line.

Because we are plotting P along the horizontal axis, to the right of this line P is decreasing, so trajectories must move to the left. To the left of the isocline P is increasing, resulting in trajectories that move to the right. On the line itself, P is (temporarily) constant. The arrows in the regions of the portrait indicate these directions.

If we set dC/dt to zero we get

$$C = 1 - \frac{n}{b}P$$

which gives a second isocline, shown in Figure 11.2.

Because C is plotted on the vertical axis, the arrows on the phase portrait are indicating the vertical motion of trajectories. These two lines may or may not intersect. If they do intersect, we have an equilibrium point, shown in Figure 11.3. If not, we see something like Figure 11.4.

Because of the different roles of the two lines in determining direction of motion, we actually get four scenarios, depending on the constants in our equations, shown in Figure 11.5.

For each of these possibilities we only need to think about trajectories that begin inside the rectangle determined by the two limiting

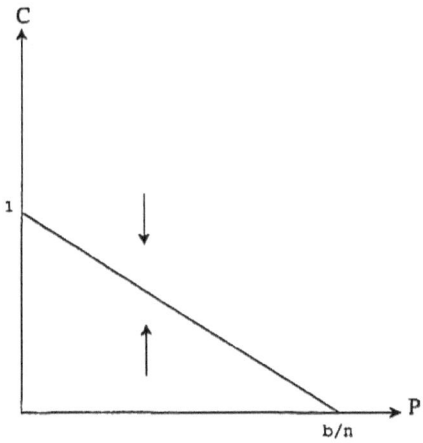

Figure 11.2: $dC/dt = 0$ isocline.

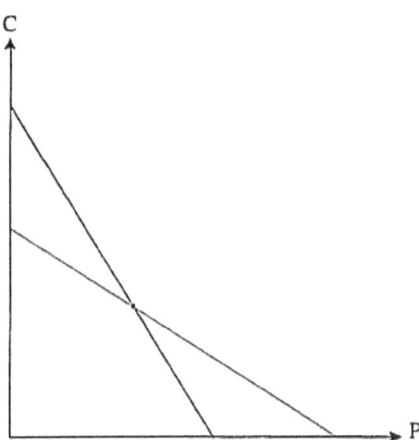

Figure 11.3: An equilibrium point where isoclines cross.

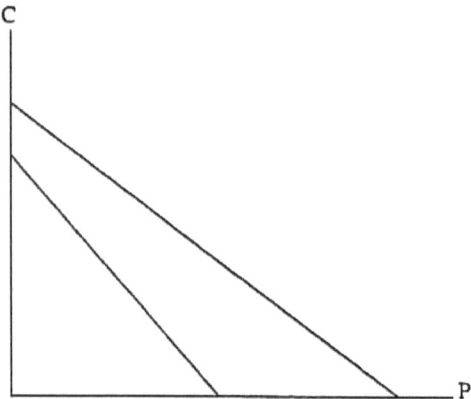

Figure 11.4: No equilibrium point inside the quadrant.

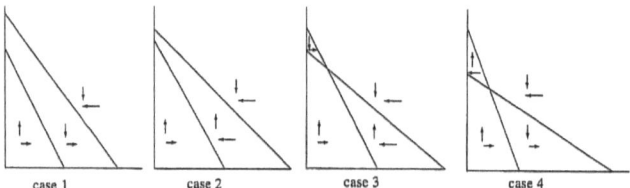

Figure 11.5: All four possible options.

populations, because initial populations above the carrying capacity are unlikely. So, for the first scenario in Figure 11.5, we are looking at a box containing three regions whose long term behavior may be different. We need to look at each of these separately. They are labelled A, B, and C in Figure 11.6.

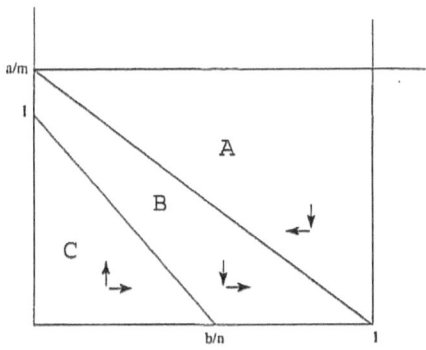

Figure 11.6: Isoclines for first case.

In the region labelled A, all trajectories are moving downwards and to the left. It is easy to see that they will eventually have to enter region B. In region C, all trajectories are moving upward and to the right. It is easy to see they will also enter region B eventually. In region B, however, the trajectories are moving down and to the right. They cannot enter region C because when they get to the isocline all downward motion stops and they are forced to the right, back into B. For a similar reason, they can't enter A either.

Within B, they move forever downward and to the right. The catfish are dying out and the perch are approaching their limiting population. All roads lead to the fixed point at the intersection of the horizontal axis and the higher isocline. In this scenario, the perch will outcompete the catfish and drive it to extinction, no matter what the relative populations are when the perch is introduced.

This kind of analysis can be done for each of the four scenarios. The reader will find, after having completed the exercise, that in the first two of the scenarios one species is always forced to extinction. In the third scenario both species co-exist and come to equilibrium. In the last scenario, the final outcome depends on the original populations of both species.

Our example illustrates several important ideas. First of all, we see the importance of thinking through simple algebraic consequences of our model, as they apply to the phase portrait for it.

Second, when analyzing a complex system using time series output from the computer, one must try a variety of initial conditions. If the phase portrait lurking behind the numerical computation is like the fourth one above, there will be very different conclusions depending on what the populations were at the beginning of the calculation.

Third, we see how very important it is to have an idea of what the constants are in our equations. A slight change in the constant might have enormous long term effects on the system, as illustrated in Figure 11.7.

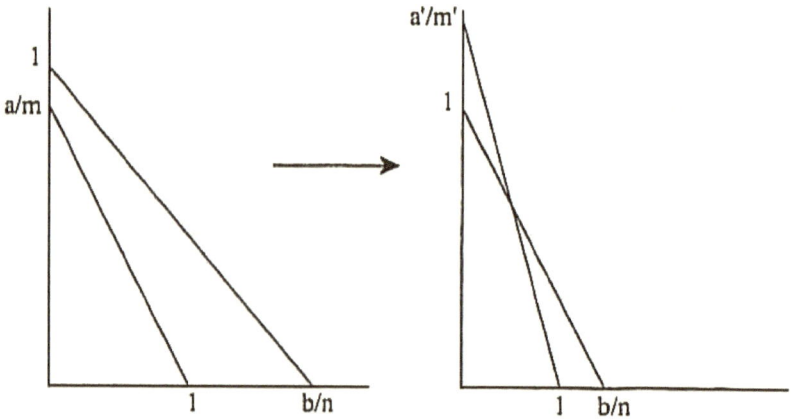

Figure 11.7: A change in a/m.

Fourth, we see that the mathematics gives us enormous insight into natural systems. To the extent that we believe our equations are a fair representation of the relationships between organisms, the phase portrait analysis vividly portrays the range of possible outcomes and the factors upon which each outcome depends. This is useful information for the purpose of both predicting outcomes in an ecosystem and also for interfering with those outcomes. Any modification in behavior of one of the two organisms in our example will affect one or more of the constants involved. Human behavior could also be added to this example, such as fishing patterns that favor catching one species over the other. In fact, human fishing patterns on the lake are tailored to the catching of perch, which is highly exportable, rather than catfish. The reader is invited to investigate how such behavior would alter the constants in the above equation so that the Bagrus catfish, although low in population, may be prevented from extinction. These models, and their interpretation, are the only predictive guides we have to intervention in the large, complex systems that routinely arise in ecology, physiology, and epidemiology.

For your consideration

Question 1:

In which of the four possible scenarios does a species always become extinct? In which case do both species always survive? Draw a picture of a trajectory in this case. What are the constraints on a, b, n, m? How would you explain the physical meaning of the parameters in these equations? What might affect them?

Question 2:

One case has different outcomes depending on the initial position. Which is it? Draw the phase plane and indication which regions lead to which long term behaviors of the population. What are the implications of this in term of ecological intervention? The general term for this sort of phenomenon is "population dependent effects", because the long term behavior of the system is dependent on the starting populations (unlike the other three scenarios).

Question 3:

There are two scenarios that lead to extinction of one species or the other. As the parameters change, one morphs into the other. Is it possible to perturb the parameters in such a way that there is no time at which coexistence is possible? Explain the implications of your answer in terms of ecological intervention.

Chapter 12

Species Formation

Five hundred distinct species of cichlid fish are endemic to Lake Victoria. The word "endemic" means that they are found there and nowhere else. In the case of Lake Victoria cichlids, this is not strictly true as they are a popular freshwater aquarium fish in tanks around the world. But their only natural habitat is Lake Victoria. Other nearby tropical lakes boast a similar diversity of (different) cichlid species. Geologists believe that Lake Victoria suffered a drought about 14,700 years ago, drying out completely. As it refilled, some ancestors of the current cichlids must have arrived. If the current diversity of the lake were derived from a single originating species, then the rate of evolution of individual cichlid species would be about one new species every 30 years.

Biologists dispute the 14,700 year geological estimate, based on DNA evidence in the cichlid population. They theorize that the lake did not dry up completely, preserving the cichlids in a series of disconnected ponds. If we disregard the drought of 14,700 years ago, the previous period of complete dryness would be about 100,000 years ago. This estimate would result in a new species appearing approximately every 200 years. Furthermore, the hypothesis of preservation of the species in many disconnected ponds over an extended period provides a mechanism for evolution: the process of isolation and reintroduction of species.

When a population of species is cut into two parts that can no longer interbreed, the genetic makeup of the two separate cohorts will slowly drift apart. Mutations are random, and a sequence of mutations in the DNA of a population causes the genetic makeup of that group to vary from generation to generation. A mathematician would say that the mutations are causing the genetic composition to take a "random walk" , with slight changes accumulating to cause, in the end, a real divergence of genetic makeup between the two cohorts. Of course, this process occurs in response to changes in the rest of the ecosystem. In the case of fish in two newly separated ponds, all the other species in the pond undergo the same process. The population of all organisms in the pond is a linked ecosystem coevolving.

Ultimately enough difference accumulates between the two fish populations that they can no longer interbreed. It may take only a small difference: a missing visual cue, a slight change in breeding cycle. If the two populations are then mixed, the two populations may refuse to hybridize at all, remaining separate breeding pools. Sharing the same habitat, they now compete for resources. We can use the model from the last chapter to investigate the claim that isolation followed by reintroduction is a viable way of developing new species.

As we saw in the last chapter, it could easily be the case that one species is driven to extinction. In fact, in 3 of the 4 cases we looked at, this was indeed the case. In other words, if the parameters of the equations for the competition model were chosen at random, extinction of one of the two populations would be the most likely scenario. Of course, it is not the case that the parameters are random in the case of isolation and reintroduction. Rather, a few random mutations have occurred that may or may not alter each of the parameters in the competition equation.

Many mutations are inconsequential to survival. Most of the features we use to distinguish ourselves from each other as humans are not really important features from an evolutionary standpoint. So perhaps the first question we should ask about our two species of fish is what the competition equations will look like if the two species are identical in every way that might affect the four parameters in those equations. In this case we get two identical equations:

$$X' = aX(1 - X - Y) = aX(1 - X) - aXY$$

$$Y' = aY(1 - Y - X) = aY(1 - Y) - aXY$$

The assumption here is that both X and Y were growing according to the logistic equation in separate compartments until reintroduced. At this point they share resources. We have choice as to whether we want to consider the resource available to the combined population as still equal to 1 or whether we assume it is now doubled. In any case, if we assume both populations start at the same size, Figure 12.1 shows what happens.

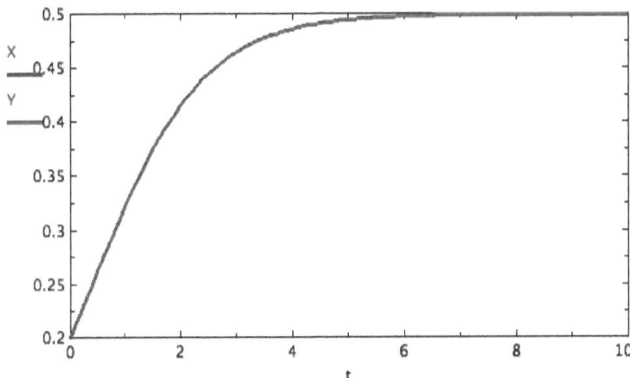

Figure 12.1: Both X and Y starting at .2 with $a = 1$.

On the other hand, it is possible that one of the isolated habitats was smaller than the other, in which case one population (say X) starts out with fewer members than the other. In that case, what happens? Figure 12.2 gives an example.

As we can see, when neither nature nor mathematics can distinguish between the populations, both persist. The difference in initial values, however, determines the long range proportions of each in the combined system.

Now that we know what happens if there is no significant mutation, let us see what happens if the populations diversify enough to affect either growth rates or the effect of competition. First let us suppose that a mutation improves the growth rate of one of the

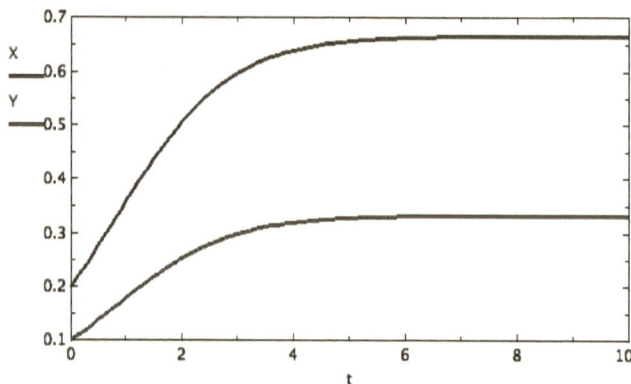

Figure 12.2: Different starting values for X and Y with $a = 1$.

two populations, without affecting the way they share the resources. This would yield two equations of the form:

$$X' = aX(1 - X - Y) = aX(1 - X) - aXY$$
$$Y' = bY(1 - Y - X) = bY(1 - Y) - bXY$$

Now what happens? We can see from the following runs with $a = 1$, $b = 2$, and various starting values of X and Y in figure 12.3.

As we see, coexistence is still likely. Now we can ask what sort of effect a mutation might have on the way the two species interact. If the two species develop slightly different food preferences, this effectively reduces the competition between them. Keeping the basic growth rate the same, we then have two equations that will look like this:

$$X' = aX(1 - X) - mXY$$
$$Y' = aY(1 - Y) - nXY$$

Comparing this with the system for identical species, our hypothesis would require that m and n both be slightly less than a. In other words, species Y has a less detrimental effect on the habitat of X than does X itself, and vice-versa. The analysis we did on phase portraits in the last chapter tells us which situation we see in this case. The populations tend to equilibrium values. In this

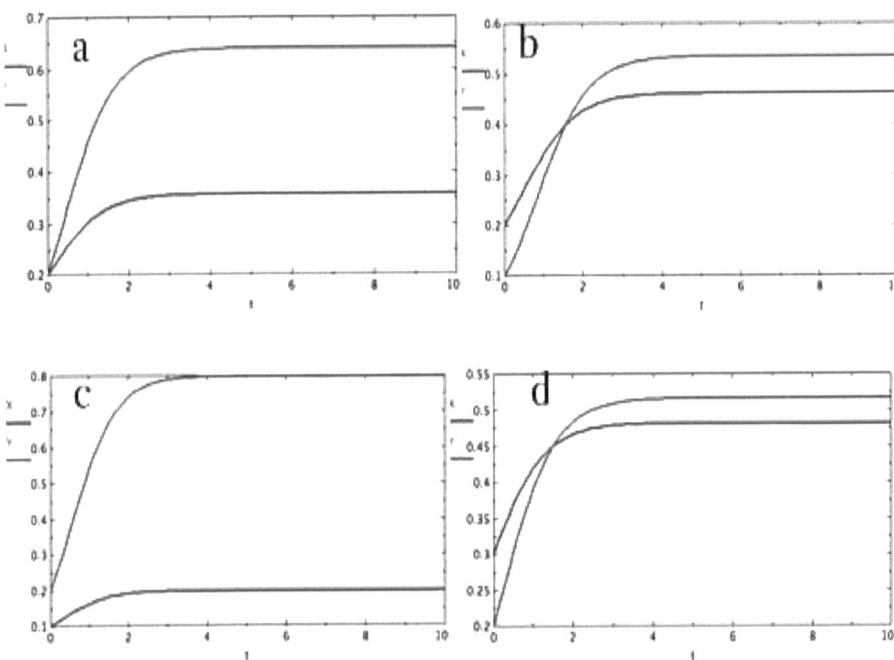

Figure 12.3: $a = 1$, $b = 2$, different starting values for X and Y.

case the equilibrium is an "attracting state", which just means that populations starting in the neighborhood of the equilibrium all end up there. Notice this is unlike the situation above with identical populations, where the equilibrium value depending on the initial populations of each type. Figure 12.4 is an example of this situation.

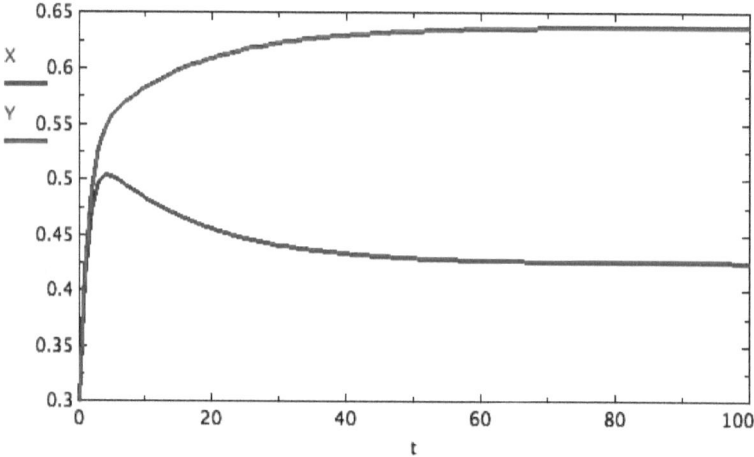

Figure 12.4: $X_0 = Y_0 = .3$, $m = .9$, $n = .85$

Ecologists have observed that even closely related species such as the cichlids of Lake Victoria and Darwin's finches often have slightly different diets, claiming that diversification of niche allows the species to cohabit successfully. Even zooplankton have been observed to differentiate themselves via food preference. The simple competition models support the hypothesis that niche differentiation enhances the ability of species to coexist. In these models even a slight decrease in the impact each population has on the other one is enough to guarantee the presence of a stable, attracting equilibrium with populations of both species present. The models are also consistent with the hypothesis that isolation followed by reintroduction of species into contact with each other is a cause of increasing diversity of related species.

For your consideration

Of course other mutations are also possible, including ones in which one population becomes a slightly stronger competitor, perhaps catching prey more efficiently than the other. In this case either m or n may be slightly larger than the constant 'a'. What happens in this case?

Chapter 13

Some days the Bear will Eat You

In the last chapter we looked at the process of isolation and reintroduction of populations to see how the competition model explains much of why that process works to produce new species. However there are in nature examples of species that not only compete with one another but are also both prey for some predator in a higher trophic level. Lake Victoria's cichlids are certainly one example of this, as all the many species of cichlid are prey for both Nile perch and Bagrus catfish. Although the perch would not yet have been introduced at the point where previously isolated populations were reintroduced, the catfish would have been present. It is possible to combine the models for predator-prey and competition to look at models that describe this structure, as in Figure 13.1.

Looking at the simplest example with one predator and two prey, we would expect three differential equations to govern this system, as in Figure 13.2.

What equations should govern this system? For some situations, the two prey species share a food source or habitat, so we might expect that, in the absence of predator, the quantities X and Y would satisfy some kind of competition model. If we think this is the case we would add a relation between X and Y to the box model above, and express that relation as a competition:

151

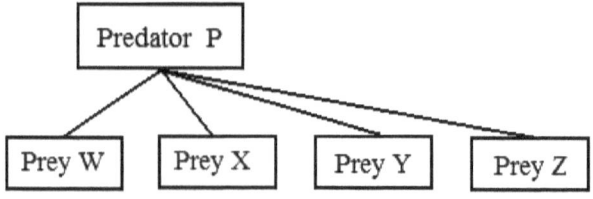

Figure 13.1: A box model for a predator with four prey species that do not interact among themselves.

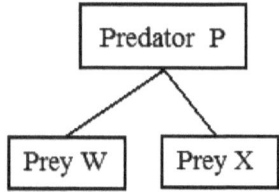

Figure 13.2: A box model for a predator with two noncompeting prey.

$$X' = aX(1 - X) - mXY$$

$$Y' = bY(1 - Y) - nXY$$

In the absence of Y, we would expect X and P to satisfy a predator prey relation:

$$P' = -kP + cPX$$

$$X' = aX(1 - X) - rXP$$

And similarly, in the absence of X, P and Y should satisfy a predator prey relation:

$$P' = -kP + dPY$$

$$Y' = bY(1 - Y) - sYP$$

Looking at these three requirements tells us that the three organisms together should satisfy:

$$P' = -kP + cPX + dPY$$

$$X' = aX(1 - X) - mXY - rXP$$

$$Y' = bY(1 - Y) - nXY - sYP$$

It is worth pausing here to notice that this analysis of our system, consisting of articulating hypotheses about pairs of organisms and then stating these mathematically, is the right way to approach complex systems. It is the only way to guarantee that the simpler components of the system behave correctly when the other parts are absent. We have to guarantee this in case one of the organisms goes to extinction in our complex model, which we will see is sometimes the case.

See Figure 13.3 for some possible outcomes for this system, with various parameter choices. Although each predator-prey pair has a coexistent equilibrium with the second prey species is missing, when all three are put together one of the two prey species may be driven to extinction.

If we want to know what possible equilibria look like, we can solve for them using these three equations:

$$0 = -kP + cPX + cPY = P(-k + cX + dY)$$

$$0 = aX(1 - X) - mXY - rXP = X(a(1 - X) - mY - rP)$$

$$0 = bY(1 - Y) - nXY - sYP = Y(b(1 - Y) - nX - sP)$$

We certainly get equilibrium at extinction ($X = Y = P = 0$) and we also get other equilibria when we set one or two of the quantities equal to zero. Here are all possible equilibria triples where one or more quantities is zero:

1. $X = Y = P = 0$

2. $X = P = 0, Y = 1$

3. $Y = P = 0, X = 1$

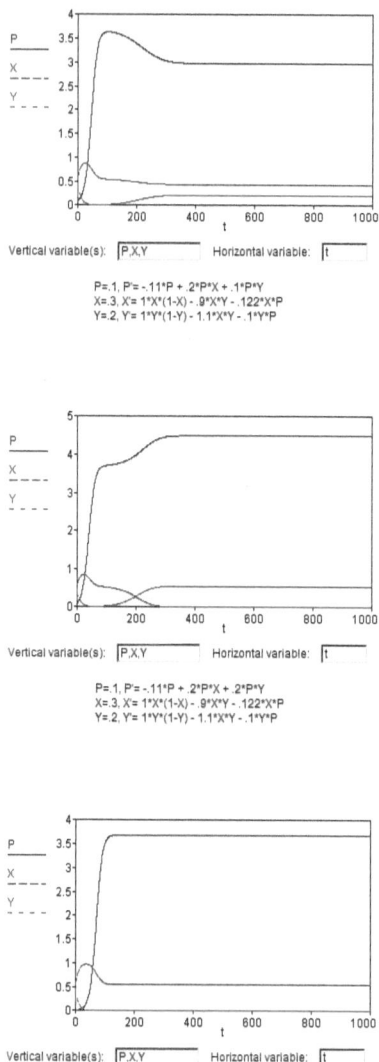

Figure 13.3: Various outcomes for a predator with competing prey with different parameters: coexistence or extinction of one of the two prey species.

4. $X = 0, Y = k/d, P = b/s(1 - k/d)$

5. $Y = 0, X = k/c, P = a/r(1 - k/c)$

6. $P = 0, X = (a/m - 1)/(a/m - n/b), Y = (b/n - 1)/(b/n - m/a)$, valid where positive

In all of these cases, one or more populations is extinct. However there is a possibility of coexistence when X, Y, and P are all nonzero solutions of:

$$0 = (-k + cX + dY)$$

$$0 = (a(1 - X) - mY - rP)$$

$$0 = (b(1 - Y) - nX - sP)$$

These three linear equations can be solved by software or brute force. For some parameter choices the solution will yield three positive numbers for X, Y, and P, which are biologically valid solutions. However, for some parameter choices the equilibrium will be unstable and therefore will never appear in simulations. For other choices the equilibrium will indeed be stable.

This situation raises the following question. Could it be that the presence of a top predator is a prerequisite to species diversity in some cases? In other words, is it possible that nature could have species that satisfy these criteria:

1. When left to themselves, one species always forces the other to extinction

2. But will coexist in the presence of a suitable joint predator?

We can test this question with our model. Criterion 1 means that, according to the analysis in chapter 11, either b/n or a/m is less than 1 and the other is greater. Suppose b/n is less than one and a/m is greater than one. Then X survives and Y becomes extinct. Here is an example of such a pair:

$$X' = X(1 - X) - .9XY$$

$$Y' = Y(1 - Y) - (1.1)XY$$

We can think of this situation as two recently divergent species of cichlid, one of which will surely outcompete the other upon reintroduction. Then we can introduce a top predator, P, with preferential feeding habits. The equations below represent such a predator, having a slightly more detrimental effect on the population of X than on Y.

$$P' = -.11P + .2PX + .1PY$$

$$X' = 1X(1 - X) - .9XY - .122XP$$

$$Y' = 1Y(1 - Y) - 1.1XY - .1YP$$

We can see the contrast in outcomes depending on whether there is predator present, as in Figure 13.5, or not, as in Figure 13.4.

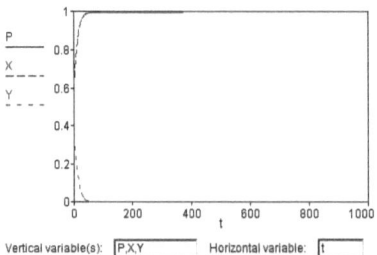

Figure 13.4: Outcome with no predator present.

These graphs vividly illustrate the beneficial effect of such a predator on the survival of a species. If an ecosystem is behaving according to this model, removal of the predator will cause one of the prey species to go extinct.

When drawing conclusions from a model of this sort, we should worry about whether the example given above to illustrate a particular principle is just a very special case. In other words, if we vary the parameters in these equations a bit (but still satisfying the

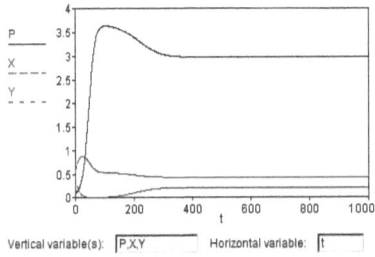

Figure 13.5: Outcome with predator present.

criteria we wanted), will the phenomenon disappear? This question is one of "robustness", which refers to the ability of a system to keep the same qualitative outcome under small perturbations of parameters. There are 9 parameters in the above equation, which represent the 9 constants that were chosen to illustrate this phenomenon. Two key constants represent the preference of the predator for consuming X over Y. These are the coefficient of PX in the first equation and the coefficient of PX in the second equation.

The coefficient of PX in the second equation can be varied, and we can see the response of the system as it goes to .2 and higher in figure 13.6.

Note that, although the different in equilibrium values for the two prey species increases, they both persist. Yet in all of these cases, if the predator is removed, Y goes extinct. The contrast can be quite dramatic, as in this case where Y goes from being the predominant prey to extinction as in figure 13.7.

However if the coefficient of PX in the second equation drops to .1 so that the predator has no more detrimental effect on X than on Y, Y goes extinct, as in figure 13.8.

We would expect Y to go extinct in this case because we have set up equations where X naturally drives Y to extinction. But the large range in which X and Y coexist as we change this parameter tells us that the system is robust with respect to that parameter. For a full study of robustness, we would test all of the parameters in this way. Robustness of the system gives us confidence that a real system would behave this way, because a real system is subject

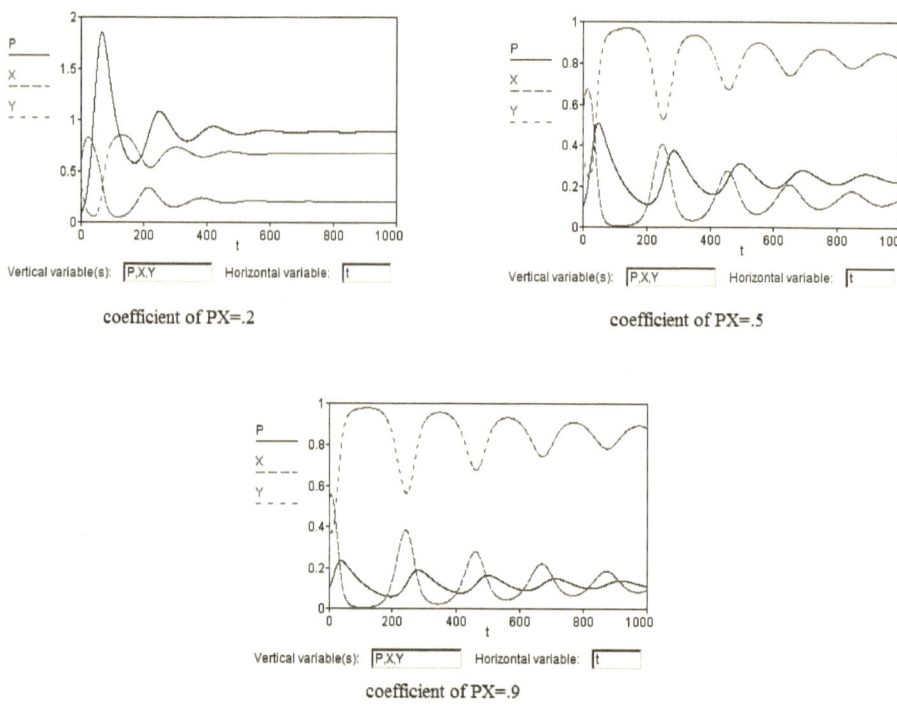

coefficient of PX=.2

coefficient of PX=.5

coefficient of PX=.9

Figure 13.6: Varying coefficients of PX in the second equation of the model.

to small random perturbations. Our model would represent these as small perturbations of the parameters in the equations, and we want our qualitative conclusions to remain valid under these sorts of small changes.

One way to think of this collection of models is as support for a hypothesis. In this case the hypothesis is "The existence of a top predator can make coexistence possible for species that otherwise could not coexist". We restated this in mathematical form, "If the hypothesis is true we should be able to find a model where two prey species coexist in the presence of a predator and do not coexist when the predator is removed". We tested the hypothesis by constructing equations representing two species that would not coexistence in

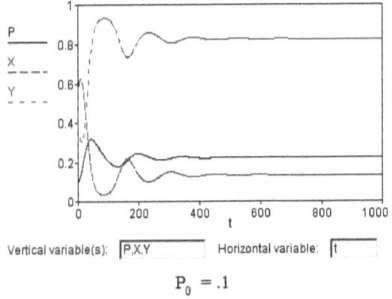
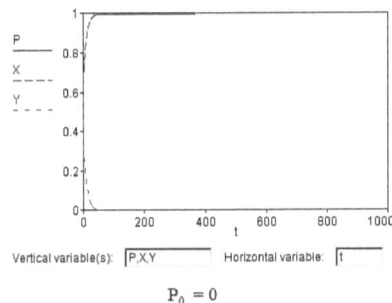

$$P_0 = .1 \qquad\qquad\qquad P_0 = 0$$

Figure 13.7: On the left, Y is the highest function yet on the right it goes to zero with the predator removed from the system.

competition, plus a predator that interacted with each of them. We found parameters that satisfied both requirements, and we showed that the system is robust under small changes in these parameters (actually we only demonstrated how to do this with one of them). Now we can be far more confident that our hypothesis is correct, because when we quantify it properly it holds true. This example is an excellent illustration of how models are used to test concepts, even when the real parameters are not known.

Ecological concepts are very difficult to test in nature, because we don't care to experiment on our own ecosystem. But sometimes experiments just happen anyway. There is a wonderful account, described by E.O.Wilson in "The Diversity of Life", (Wilson, 1999) of the removal of sea otters from the Pacific coast of California, Oregon and Washington in the early part of this century. The sea otters were hunted nearly to extinction for their warm pelts, removing a top predator from the ecosystem of giant kelp beds up and down the coast. These ecosystems are abundant and rich, far too complex to model each species. However they are all based on the abundant algae in one way or another. Furthermore, there is one species, the sea urchin, which is only eaten by sea otters. It has no other predator and is not linked to the other species, except through competition for algae. So we could lump a large complicated group

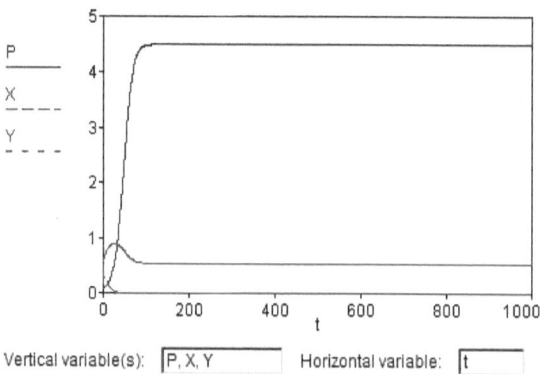

Figure 13.8: For sufficiently small coefficient of PX in equation 2, Y goes to extinction.

of interacting species into one biomass (labeled below "everything else"), which is in competition with sea urchins. Both of these lose biomass to the top predator, the sea otter. Our box model looks like the one we have been discussing in this chapter, as seen in Figure 13.9 (without the arrows representing competition).

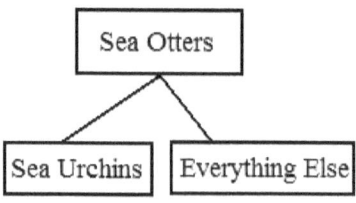

Figure 13.9: Box Model with Sea Otter, Sea Urchins, and "Everything Else".

The discussion in this chapter would lead us to believe that, with the otter removed, there are three possibilities. It is possible that both prey boxes continue to coexist, or that one is driven to extinction.

What actually happened? Not so long after the otters were removed the diversity of the kelp forests disappeared, and they were described by witnesses as "sea urchin deserts". The lowly urchin managed to outcompete its complex diverse competition. Some years later when the otter was reintroduced, the diversity of the kelp forests returned. Rarely do we see an ecology experiment on such a grand scale. For a convincing explanation of why the simple ecosytem consisting of algae/urchin might be the winner in competition with a complex, highly developed trophic web, read "Why Big Fierce Animals Are Rare" by Paul Colinvaux (1979). The answer he gives is, although free of equations, a deeply mathematical discussion.

The concepts of a top predator making it possible for other species to exist plays out not only in lake Victoria, and other water areas, but also on land. Large animals have a big effect on their environment, irrespectively of what the environment is. These concepts can be examined therefore through the relationship between large African Herbivores and their habitat. For general information and conceptual information with data, see Birkett & Stevens-Wood (2005), Birkett (2002), Cromsigt *et al* (2002), Mackey *et al* (2006), Nicholls *et al* (1996), and Koppel & Prins (1998). In addition to the competition among herbivores described in these articles, one might also explore the effect of introducing or removing a top predator.

For your consideration

Question 1:

Is it possible to have three competing species, X, Y, Z that always results in coexistence of X and Y and extinction of Z, yet in the presence of a common predator, X and Z survive and Y is driven to extinction? If you can figure out a likely reason why such a thing might happen, you can test it by embodying your reason in a set of four equations. Then you can test to see if it is possible.

Question 2:

In regard to question 1, you might think the competition equation for X looks like this:

$$X' = aX(1 - X) - bXYZ$$

There is a good reason why this is not the case. Can you figure out what it is?

Chapter 14

Eichhornia Fecunda

Human beings are very good at moving organisms around the planet. We move food crops, diseases, ornamental plants and pets. The same fish tank that contains a species of fish that may already be extinct in Africa may also contain a plant that has come from afar to make its home in Africa. It seems to be human nature to move organisms around, but sometimes we create problems for ourselves by doing so.

The Brazilian floating plant *Eichhornia crassipes*, is not especially remarkable. It is about six inches tall, produces both sexually and asexually through runners and buds, and produces a pretty lavender blue flower that explains its common name of "water hyacinth".

One day, it is uncertain when or how, someone noticed how lovely and hardy the water hyacinth colony was and brought it home. Several live specimens came to the United States in the 1880's, to a fair where they were bought by the hobbyists and bred for resale. From those few fledgling plants, many daughters were produced which spread from pond to pond, lake to lake, state to state, and nation to nation. In the absence of any natural predators, and with the encouragement of gardeners entranced by its beauty, the water hyacinth spread until it became naturalized throughout most of the southern states. It spread so quickly and successfully that people became wary; anything that outpaces humans seems to come under scrutiny from biologists, state officials, and government agencies. Each of these parties discovered that, while water hyacinth is beautiful to look at, it is extremely hard to eradicate.

Perhaps it spread from a garden pond to a municipal sewer system; water hyacinth thrives in water polluted with excess nutrients. Then it may have clogged intake pipes to treatment plants, or perhaps multiplied aggressively in holding ponds. This scenario was replayed in thousands of similar places across the country; word began to spread that an exotic invasive plant had arrived. Exotic invasives thrive outside of their native habitats, because conditions are perfect for their growth and they have no natural predators in their new environments. They also outcompete native species, whose ecosystem functions people tend to disregard until they are gone. Water hyacinth accomplished all these deeds - the innocuous, lovely plant from South America crept into every corner of the southern US, and continues to spread, despite the fact it is illegal to buy, sell, or transport it; you face a hefty fine and/or time in jail if you do so. Despite the threat, it is still sold in nurseries, or traded privately - it is hard to resist something so attractive, and so easy to grow.

Water hyacinth is not just a problem in the United States. In fact, it is considered to be the worst aquatic plant in the world; from India, Bangladesh, Pakistan, the Philippines, Thailand, Malaysia and Australasia to most of subtropical Africa it has spread ceaselessly. In Eastern Africa, where the problem seems to be particularly intense, water hyacinth is classified as a noxious weed, and its culture is illegal in some places. African lakes have a long history of being invaded - not by warring factions, but by plants. The lakes are usually large, and have many niches for different types of plants; colonists also had their own ideas of appropriate flora, and tended to import species to make their new environs feel more like home. Home, apparently, was full of water hyacinths. The lakes are also located in a subtropical climate, which worldwide is just behind the tropics in terms of biodiversity - the number of species which can comfortably call it home.

Lake Victoria provides a perfect case study for water hyacinth invasion. Unlike the early spread in the U.S., the Lake Victoria spread happened much more recently, and good records of the infestation exist. It seems that water hyacinth first came down the Kagera River, which empties into Lake Victoria, in large floating mats dislodged from upstream. Some may also have come from ornamental ponds around Nairobi, Kenya as well; water hyacinth was kept in ponds at least as far back as 1957, and it made its first documented appearance in Lake Victoria in 1959. Water hyacinth is still flowing down the Kagera River and into Lake Victoria; one study estimated that approximately 0.2-1.5 ha enter the lake daily. In addition, the study calculated 1% growth per day by already present plants, meaning that 64 ha along the shoreline could generate 0.64 ha/day of new growth in a single day. In the Congo River, another area where water hyacinth has made its home, two plants were observed to produce over 1200 daughters in the space of four months. This rapid rate of growth accounts for the rate of spread in Lake Victoria; in 1991, water hyacinth was observed to cover 1% of the 27,000 square miles of lake. The estimate as of this writing in 2000 is 3% (10,000 ha), and is increasing exponentially; if two plants can produce 1200 daughters, and each of those 1200 daughters can produce at least 600 plants apiece in perpetuity. 80% of the Ugandan shoreline is infested to at least ten meters out, even beyond, and areas such as the Mwanza Gulf are sometimes blocked over their entire width - in this case, almost five kilometers. Away from the shoreline, free-floating dense mats of water hyacinth nearly 600 ha in area sometimes are blown into bays on the North side of the lake by southerly winds during the early part of this century, blocking traffic for days, until they are blown back out again by a north wind.

One of the most important factors accounting for the rapid spread of the water hyacinth is that it has evolved to reproduce and grow maximally in many conditions. The plant is self-pollinated, with the assistance of insects, but also can (and usually does) reproduce vegetatively, through runners like the ones described in the Brazilian plant above. If it can reproduce sexually, it will set seed, which sinks to the bottom, lodging in the sediments, and can last for up to 30 years - until conditions are right for germination. The plant can survive temperatures as high as 34C, and as low as freezing; the leaves can be killed by frost, but in order to kill the entire plant and halt growth, the rhizome tips must be completely frozen. Estimates of the time it takes for a given number of plants to double range from 8-20 days, depending on environmental conditions. One study calculated the average growth rate of the plant to be 10-12 g per square meter per day, with a maximum of 45-50 g per square meter per day. Another put it in different terms; a single plant was observed to produce 140 million daughter plants per year, enough to cover 140 ha of water with a fresh weight of 28,000 tons of plant matter.

If water hyacinth were just another lovely plant, easy to keep when it was desired, simple to divest when it was no longer wanted, this chapter would not exist. Unfortunately, the plant has proved to be more terrible than Godzilla, and larger than King Kong. It is, to many who try to make a living from Lake Victoria and such similar places, a true monster, negatively affecting navigation, fisheries, industry, water supply, and health. The dense mats block bays and the movement of ships, limit landing sites, hamper docking of ferries, and damage the engine cooling systems of boats. At Port Bell, the turnaround time for ferries has increased from 6 to 12 hours since the occurrence of water hyacinth, and over 1,000 liters of fuel are sometimes used by a ferry just to break through the snarls. The Uganda Railways Corporation is in some years spent over $12,000 per week for mechanical removal of water hyacinth at Port Bell, and Kisumu Port authorities may be forced to do the same - the plant has brought port activity to a standstill, holding up food consignments and other shipments for weeks.

Local fishermen are thwarted by mats of vegetation blocking existist from their beaches and harbors to open water. A harbor or bay can be blocked overnight when the wind blows in a green iceberg. Commercial fishing boats face the same difficulty. Aside from affecting the mobility of boats, water hyacinth affects the fishing industry by changing the habitat. The rapid growth of water hyacinth lowers oxygen levels at the shore (since respiration consumes oxygen), forcing fish to occupy less desirable habitat. Some even worry that actual changes in water level may result from the hyacinth, which has an unusually large transpiration rate. It functions as an efficient pump, moving water from the lake back into the atmosphere.

It also reduces the availability of spawning grounds, though iron-
ically the mats provide refuge for Nile Perch, another introduced
species which is outcompeting native fish. At least one study has
suggested that the hyacinth provides cover for the diminished populations of cichlid fish as well, providing the first example of potential
benefit to the lake.

Industry is affected as well. Hydropower, an important source
of energy for the surrounding nations, is severely interrupted when
the intake screens and cooling filters of turbines are clogged. This
results in higher costs for a less reliable source of energy, something
the impoverished villages and cities surrounding Lake Victoria cannot easily afford. Some estimates from early in the century claimed
that 25 million people around the lake could potentially be affected
by the plant, racking up economic losses of 150 million dollars an-
nually from an estimated 5 percent reduction in lake quality. The
reader is invited to speculate how one might measure "lake quality"
in quantitative terms.

In light of the given information, it seems at first quite clear
that something must be done to eradicate the water hyacinth, or
at least to halt its spread out of its native habitat. This attitude
is the traditional human response to change in habitat, whether
involuntary (as in this case) or voluntary. In the U.S. we can see
this attitude displayed in the everyday behavior of people who move
from the wet, green, grassy east coast to the mediterranean climate
of California and then proceed to attempt to recreate their lawns
and gardens as if the actual annual rainfall had not changed. Only
a few reactionary holdouts find their dogged determination odd.
It is this same aspect of human nature that explains why no one
runs about saying, "We must adapt to the water hyacinth!" Instead,
we compare the humble plant to Godzilla, labeling its completely
natural behavior as "invasive," a negative term usually reserved for
advancing armies of human beings.

In fact, one reason that the hyacinth can grow so rapidly in
Lake Victoria is that it is very well fed by nutrients that run into
the lake at a rate that is made all the larger for human activity.
The algae bloom described in chapter one is the forerunner of the
water hyacinth.

Agencies worldwide have attempted to address the issue, some more successfully than others.. Various herbicides such have been tried, though the effect has been minimal. It is difficult to reach all the plants with just one spraying, and multiple sprayings might negatively affect other aquatic life. The mass of dead plants from the spraying proceed to decay in situ, releasing nutrients into the water and depriving it of oxygen, which could lead to eutrophication and fish kills. Additionally, there is no guarantee that reinvasion by sunken seed will not occur months, even years later.

Another possible option is mechanical cutting and harvesting of the plant. Vaughn Co. USA manufactures a device that can harvest without detaching the seeds; Aquarius Systems US won a World Bank contract to chop 1500 ha of water hyacinth in twelve months, using two "swamp devils" and a harvester. They plan to use the harvested biomass for fertilizers and crafts. There have been no significant results from this approach, especially in shallow areas where the infestation is worst, possibly since the process has been compared to cutting away a small island. The harvesting machines can keep commercial fisheries and transportation alive, however, by clearing channels from dockside to open water. The chopping machines are also used to manage floating islands of hyacinth that sail regularly into Kisumu Port at a speed of 3 to 4 knots.

If the chemical and mechanical methods fail, there is still one other option, and it is biological. In 1997, 35,000 weevil eggs (*Neochetina bruci, Neochetia eichhornia*) were released in selected areas. There are virulent pathogens which exist specifically to target water hyacinth, and if a native strain can be developed, it will be tried. It has been suggested that the fungus *Cercospora rodmanii* may improve the success of the weevils, and that control may even be achieved by manatees, as it has in Guyana, snails, which have worked to some extent in the USA, and mites (*Tetranychus telarius*), used in Belgium. The main theme seems to be that water hyacinth is very determined, but its human foes are even more determined to eradicate it; perhaps, as some have suggested, the best solution is every solution, integrated and managed intelligently. A mathematical model could prove extremely useful in guiding this process.

Our previous models basically described the relationship between two species, which could be diagrammed as boxes with a connecting line standing for the relationship between two boxes. Each box represents a quantity that can change. Any change in a given quantity should be written in terms of the boxes connected to that quantity. In the case of Lotka Volterra or competition between species, we would have two boxes, and the relationship between them was represented by the mixed term in the differential equation. When we erased that term (the product of the two populations) we got an equation for the growth of one population in the absence of the other one.

Any model describing the effects of the water hyacinth will have a lot of boxes. The hyacinth, the weevils, the perch and cichlid are all interconnected. If we believe that the cichlid benefits from hiding in hyacinth, then the equation for change in cichlid population should include a term representing the advantage offered by the vegetation. The weevil eats hyacinth and is therefore dependent on its population, but we have to decide if the Lotka Volterra term applies in this situation of stationary prey. If not, we will have to invent another kind of term that represents the interaction better. If we include the effects of harvesting, then we have to decide what likely human behavior might be. Will it be a constant amount taken each year or will the amount vary according to some rule? If we want to take the general degradation in oxygen into account, we have to decide which species that affects and according to what rule. We have many boxes, or "compartments" in our model and each will have its own differential equation describing its rate of change, built out of its natural growth rate and all the relationships described in the diagram we build. Fortunately, once we have our equations, any decent numerical program will be able to produce solutions for us, either as time plots or as projections of the phase portrait onto any two variables we choose. Figure 14.1 is one possible configuration of boxes.

The creators of this configuration had four differential equations. For the hyacinth population, as an example, it had:

$$\frac{d\omega}{dt} = g\omega - e\omega f - k$$

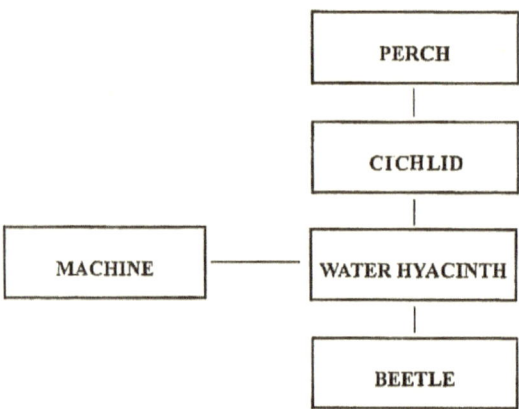

Figure 14.1: Proposed box model for hyacinth/weevil/cichlid/perch.

The first term is the natural growth rate of the hyacinth, (ω). The second term is a standard Lotka Volterra interaction with the weevil population, (e). Notice the extra term at the end. These authors used this term to model human harvesting. They are saying that the amount harvested is constant. Notice the assumptions that go with each of these terms.

The first term is saying that, in the absence of harvesting or weevils the hyacinth grows exponentially. It is the same equation we saw with algae in chapter one. Of course, there is a limiting factor to the growth of hyacinth, namely the size of the lake. If one percent of the lake is currently covered, are we still far enough from full to ignore the carrying capacity of the lake? If not, we should be using the logistic equation for this term.

The second term is saying that the predator prey relationship is a sort of hide-and-seek kind of relationship like that of fish populations. In fact, the hyacinth is fairly stationary. Perhaps, as the weevil population increases, the weevils do have to search for suitable unused spots to lay eggs, so this is a reasonable description of their activity.

The last term is claiming that human harvesting behavior is constant. Is this what people actually do? Presumably a more accurate model of human behavior could be invented.

The authors of the above model had equations for all the boxes in their diagram. Figure 14.2 and 14.3 show two examples of output for hyacinth and weevil.

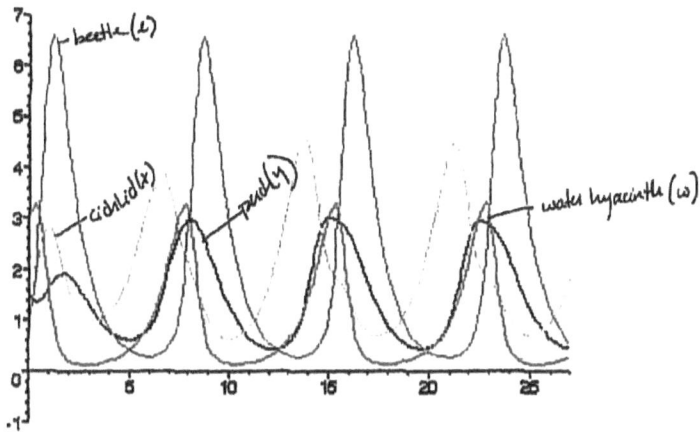

Figure 14.2: One output from proposed system.

In this example there is no harvesting and we see the classic periodicity associated to Lotka Volterra. The weevil does not eradicate the hyacinth and the hyacinth has recurrent periods of high population every five years or so. In Figure 14.3, the harvesting constant has been set very high.

Here we see the hyacinth completely controlled and the weevil population going to zero or nearly so. We must stress that, in this example, the harvesting behavior is so extreme as to be impossible for practical purposes. One of the very difficult problems for this model is deciding what constants are reasonable based on the literature. The authors of this particular model concluded that harvesters were unnecessary in the long run because of the long term behavior of the first graph above, but failed to analyze whether the recurrent blooms of hyacinth were large enough to be troublesome for human activity. They also failed to notice that overharvesting, followed by a cessation of harvesting, results in larger oscillations in the Lotka Volterra model. Pushing one population toward zero lands you on

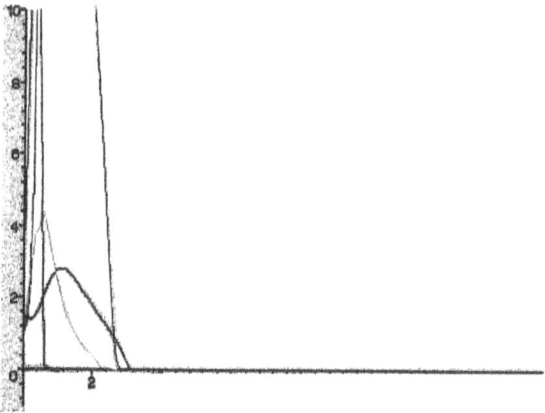

Figure 14.3: Another output from same system but with a different harvesting constant.

one of the outside loops of the phase portrait, resulting in a time series whose oscillations are farther from the equilibrium point and thus very large. A better model, more delicately analyzed, is the reader's job.

It is worth mentioning at this point that the model in this chapter was drawn from student work, and the equations and outputs (which look a little fuzzy and have some handwritten labels) are entirely due to these students working from this text and the literature. By far the hardest part of their work was figuring out what the values to give to the relevant parameters. When measuring the growth rate of hyacinth, choices of unit vary from square meters of coverage to weight, making the task of unit conversion and consistency a tough part of the problem.

In the interests of inquiry and fairness, let us forget about eradicating the poor plant and ask instead what adaptations humans might make during a periodic hyacinth boom. Surely a plant this fecund must have its uses and, if we can find them, we will surely be rewarded. It should not surprise us that many people have found ways to capitalize on the economic uses of this plant.

The harvested product has been used for silage and soil compost in Uganda, as well as for weaving twine and rope (Thai Trading Company offers water hyacinth cord on its website). There are two biogas plants which make use of it as fuel, and on a small scale it has even been used for livestock feed. It is not useful freshly harvested as fodder, since it is prickly and the seed can be spread in feces, but this is not a new dilemma. Farmers in Southeast Asia have developed a successful recipe for the feed, which calls for it to be boiled to a paste, mixed with bran, maize, and salt, and consumed within three days lest it go foul. A more specific recipe requires 40 kg water hyacinth, 15 kg rice bran, 2.5 kg fish meal, and 5 kg coconut meal. As a fuel, one kg of dried water hyacinth will produce 474 liters of biogas with 75% methane; perhaps Eastern Africa will invent the safe and cheap natural gas-powered car before the West does!

Industrially, water hyacinth has been successfully used to produce cellulose. One byproduct of the process is Xylanase, which is used in the paper industry, to upgrade animal feed, and for the clarification of fruit juice and wine. Another byproduct is a potent plant pathogen, and still another is fertilizer and soil conditioners. Best of all, this process has been established successfully on a commercial scale in India.

Still one more use involves water treatment; it turns out water hyacinth has quite an affinity for heavy metal ions, such lead, which contaminate groundwater. The San Pasqual Aquatic Treatment Facility, located in San Diego County, California, uses water hyacinth with reverse osmosis and chemicals to treat over 1,200,000 gallons of sewage per day, using the gray water byproduct to water nearby lawns. The plant is ideal for this use because it quickly takes up phosphorus and nitrogen, two of our major aquatic pollutants, it outcompetes phytoplankton, which cause fish kills, and metabolizes organic substances such as phenol, improving water transparency.

In fact, people have found water hyacinth so useful that a library search will turn up articles describing mathematical models for nutrient usage of hyacinth, with the goal of finding a fertilization regime that actually maximizes its growth rate. And on the shores of Lake Victoria we find small scale businesses arising that collect and process water hyacinth. It is made into woven furniture that is of high quality, suitable for trade and export. In some provinces, at some times, the collection of water hyacinth without a permit has been banned. This type of small scale, geographically determined harvesting could also be accounted for in a model.

From local village tribunals and international task forces surrounding the lake to the pockets of multinational agencies such as the International Monetary Fund and the World Bank, it seems everybody has a stake in the future of the Lake Victoria economy, and in preventing the spread to other nations of the life-loving water hyacinth. Nobody has yet achieved a successful method of eradication, or perfected a commercial scale safe harvesting method; but everybody seems to be working on it. As of 1999, the World Bank and the Global Environmental Facility had loaned 23 million Kenya shillings ($1US = 5 shillings); others had offered to solicit $2.5 million more for mechanical removal.

Meanwhile, nature continued to adapt. Aquarius Systems was kind enough to share with us a report from a demonstration project in Kenya. Part of the report describes the state of the hyacinth at the time chopping started in October 1999, several years after the introduction of the weevil to the lake. Evidently the weevils have considerably weakened the hyacinth, but as Lotka Volterra predicts, failed to eradicate it. Masses of floating hyacinth became colonized by secondary growth consisting of at least 23 plant species, much of which was Hippo Grass and papyrus. The personnel from this project estimated that about 30 percent of the visible plant material was hyacinth and 70 percent was other vegetation. Bushes and trees were identified on some of the floating islands and on others the weeds were more than 10 feet high. No one knows whether this secondary colonization would have happened without the introduction of weevils. It seems possible now that, for hyacinth to be used as an industrial base, the weevils themselves will have to be controlled, at least in some locations. We close this chapter with a quote from the Aquarius Systems report:

"What we can say for sure is had we not been chopping in the Port of Kisumu this winter and keeping the floating islands that arrived each day in check, it would have looked like a forest and been unusable in a matter of months."

For additional information from a scientific point of view about the Water Hyacinth and the Weevil, see Xie *et al* (2004), Wilson *et al* (2007), Njoka *et al* (2006), Heard & Winterton (2000), Cornwell *et al* (1977), both articles by Center *et al* (1999), and Albright *et al* (2004). For a weevil Hyacinth model see Wilson *et al* (2005).

For your consideration

Question 1:

Design a box model that reflects your understanding of the water hyacinth problem in Lake Victoria. Write down equations for the rates of change for each of the quantities in your model. Experiment with your model to see what it predicts as long term behavior of the weevil and hyacinth.

Modify your model to take into account the effect of human harvesting. What is your recommendation to the governments of Tanzania, Uganda, and Kenya? Can they safely discontinue harvesting water hyacinth of the weevils are present? Should they keep the harvesting machines? Write your answer as a technical report to a hypothetical international committee appointed to study this problem.

Question 2:

The effect of hyacinth on both cichlids and Nile perch fry is to provide cover, protecting them from predators. Make a new box model that attempts to model this effect for these fish. What does your model predict about perch and cichlids? Is there a benefit for the cichlids?

Question 3:

One hypothesis found in the literature is that floating mats of hyacinth have the capacity to reintroduce populations of cichlid into areas where they may have become extinct. Build a model that tries to capture this phenomenon.

As in the case with competing species, different initial conditions sometimes lead to radically different long-term outcomes. Experiment with your model to see if the long-term outcome is consistent, independent of initial conditions.

Chapter 15

The Hyacinth and the Weevil

An alert reader of the last chapter might be wondering if the assumptions that lead to the usual predator prey equations might be inappropriate for a situation where the prey is a plant. After all, plants don't run away. They behave more like a renewable resource than an elusive quarry. In evolutionary response, the water hyacinth weevil doesn't need to spend a lot of personal resources searching for its prey and thus might not be subject to the same dynamic as the predator prey equation predicts. In particular, suppose the weevil and hyacinth obeyed the standard (damped) equations:

$$
\begin{aligned}
N &= \text{hyacinth} \\
W &= \text{weevil} \\
N' &= rN(1 - N) - cNW \\
W' &= -gW + cNW
\end{aligned}
$$

Then the second equation, $W' = -gW + cNW$, says that at low concentrations of hyacinth, the weevil is dying off. That is, if cNW is less than gW, the growth rate W' is negative. Perhaps this is not a good assumption, in that only a small amount of hyacinth might be necessary to accommodate the reproduction needs of weevils.

This system might also be subject to a nonlinear predator functional response such as the one we saw in chapter 10. Such a response might further be supposed to have a large effect on the hyacinth population but less of an effect on the weevil population. That is, the death rate of hyacinth would be very much affected by the rate of consumption by weevils, but the birth rate of weevils might actually depend more on the amount of hyacinth available rather than the rate at which it is eaten. This situation would occur if weevils quickly achieve their maximal feeding rate.

Such assumptions are built into a model studied by John Ross Wilson (Wilson *et al* (2005) and (2007)). This model replaces the equation for weevil growth by a modified logistic equation whose carrying capacity is given as a proportion of the water hyacinth density. Its equation for plant growth includes a nonlinear predator response like the one in chapter 10 of this text. Furthermore the model includes a slight steady influx of hyacinth into the system, which we will set to zero for the purposes of this chapter. The units are all in terms of water hyacinth and weevil densities (grams of either hyacinth or weevil per square meter). The model looks like this, with N representing hyacinth and W the weevil:

$$N' = rN(1 - N/k) - c(N/(N + h))W$$
$$W' = sW(1 - jW/N)$$

Wilson's model has three advantages over the standard predator prey model we explored in this context in the last chapter. First of all, its assumptions seem to fit the situation better. Second, most of the constants that appear in these equations have been measured to some extent. Third, for the purposes of this text, Wilson's model exhibits some very interesting behavior as its parameters are varied. The kinds of behaviors we see in this model are actually quite typical of complex systems, and every modeler needs to be aware of the possibility that these might occur.

Let us begin by looking at how one might measure the constants in these equations. When there are no weevils and no influx of water hyacinth into the system, N obeys the simple logistic equation:

$$N' = rN(1 - N/k)$$

At small values of N, this is approximately $N' = rN$. So r is just the natural growth rate of N with unlimited resources, given as gain per gram of water hyacinth per day in this case. We could measure this by growing water hyacinth in a tank. Of course r will vary depending on the nutrients available and so really should be regarded as a parameter that takes values in some interval. Wilson uses $r = .08$ grams of growth per gram of hyacinth per day.

At large enough values of N growth will stop. Of course plants never actually stop growing; what happens is that the rate of death catches up with the growth rate, due to overcrowding, nutrient limits, or self-shading. When this happens the density of the hyacinth mat remains constant. One could obtain this constant by simply measuring the grams of hyacinth per square meter in a mature hyacinth mat. Again, some variation will be possible. Wilson gives 70 kg per square meter as this value. Notice the units.

The term $c(N/(N + h))W$ represents consumption of hyacinth by weevil with a nonlinear predator response. W is just the density of weevils (in grams of weevil per square meter). The response term $(N/(N + h))$ goes from zero (at $N = 0$) to one (as N approaches infinity). So if N is quite large (which happens when there are few weevils and relatively large amounts of plant material) then the constant c just represents the rate at which weevils eat hyacinth when the plant is relatively abundant. This constant can be measured in a controlled tank or observed in nature. Wilson gives it as 4 grams of weevil growth per gram of weevil per day. Notice that the term $N/(N + h)$ is dimensionless (because the numerator and denominator have the same units).

The constant h determines how quickly hyacinth consumption rises from zero to maximum as hyacinth density increases. When $N = h$, this rate (given by $N/(N + h)$) is at half of its maximum. $(h/(h+h) = \frac{1}{2}$ which is half of 1). Wilson gives the value of h as 200 grams of hyacinth per square meter (notice this is the same unit as for N). To see if this is small or large, we should compare it to the maximum value N could take, which is 70 kg or 70,000 grams per square meter. 200 grams per square meter is relatively small, so in this model the weevil will achieve its maximum rate of consumption fairly quickly. This parameter, however, is the hardest to estimate from data or measure directly.

The equation for weevil growth is just a modification of the logistic equation:

$$W' = sW(1 - jW/N)$$

When the amount N of hyacinth is quite large, this equation is approximately $W' = sW$. Thus s represents the growth rate of weevil in the presence of abundant resource. This constant can be measured in a tank or perhaps in the wild. Wilson gives it as .06 grams of weevil growth per gram of weevil per day.

As N diminishes or W grows, the effect of limited resources appears. For a given ratio of N to W the population grows, dies, or stays the same. In other words, at equilibrium when there is no growth or death, $W' = 0$. Solving this for the nonzero solution we get

$N/W = j$

The constant j is thus the critical ratio at which growth stops. This is a constant that can be measured by a controlled experiment in a tank or inferred perhaps from observations in the wild. Wilson gives 50 grams hyacinth per gram of weevil for this constant.

Putting all of these together gives a set of equations:

$$N' = (.08)N(1 - N/(70,000)) - 4(N/(N + (h)))W$$

$$W' = (.06)W(1 - (50)W/N)$$

Wilson uses $h = 200$, but because this parameter is the one that is hardest to measure experimentally we will now investigate the behavior of solutions to these equations as h is varied. Remember that a large h causes the function $N/(N+h)$ to rise slowly, reflecting inefficiency in the weevil's search for food. A low h causes the function to rise to its maximum quickly, reflecting greater efficiency of the weevil's ability to consume hyacinth.

Let us do a numerical experiment in which we drop h from 2000 down to 200. See figure 15.1 for four runs at $h = 2000, 700, 550, 200$.

What do we notice? In the first 2 runs the system goes to a stable equilibrium value. If you test intermediate values between 2000 and 700 you will see that the system (with these starting values) always reaches a stable equilibrium, although the location of that

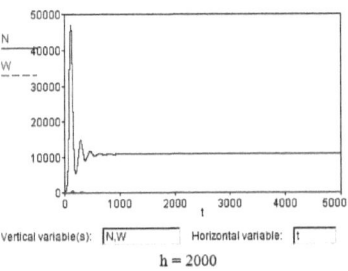

h = 2000

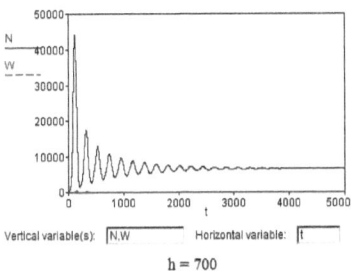

h = 700

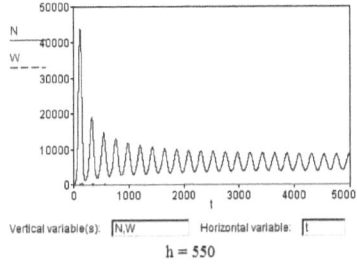

h = 550

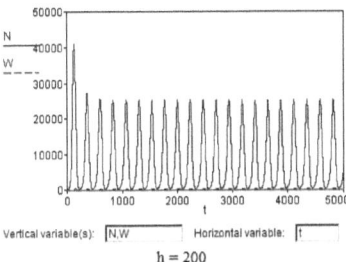

h = 200

Figure 15.1: A series of runs of the system with different choices of parameter h.

equilibrium changes. At $h = 550$ something completely different is happening. The system settles down but never reaches equilibrium. Instead it oscillates forever with a fixed period and amplitude. Notice the difference between this behavior and ordinary undamped predator-prey, which oscillates forever but with an amplitude that depends on initial populations. By contrast, here we have an oscillating system that always seeks the same amplitude for a given choice of parameters. You should try this out on the software of your choice to check that no matter what initial values you put in, the system tends toward the same oscillating behavior with an amplitude that is characteristic of the parameters you chose for the system. This phenomenon is called a "stable limit cycle", and is a common occurrence for systems of ordinary differential equations.

Somewhere between $h = 700$ and $h = 550$ a change occurs. If you test intermediate values as you lower h from 700 to 550 you see oscillations that persist for increasing periods of time until some particular value h_b where the oscillations persist forever. After this value the amplitude of the oscillations increases. A "bifurcation" is a value of a varying parameter where the system suddenly does something qualitatively different from before. The special value h_b is said to be the location of a "Hopf bifurcation". A Hopf bifurcation is a value of a varying parameter where the system changes from having a stable equilibrium to displaying a stable limit cycle.

It is easier to visualize this change in the phase portrait. Let us run a series of outputs as before with h decreasing from 700 to 550 at intervals of 50 (figure 15.2). You can see the location of the stable limit move as the parameter is changed, with the limit cycle appearing and subsequently growing.

If we were to calculate the equilibria of the system at $h = 200$, we would find there is one inside the stable limit cycle. Figure 15.3 shows what happens if we start a run at $h = 200$ with values inside the loop of the limit cycle.

As you see, the stable equilibrium that persisted for a long time has become unstable when the limit cycle appears. There are mathematical ways of checking for the exact choice of h at which stability of the equilibrium is lost. (The advanced student is encouraged to look up the Hartman-Grobman theorem at this point.) This change

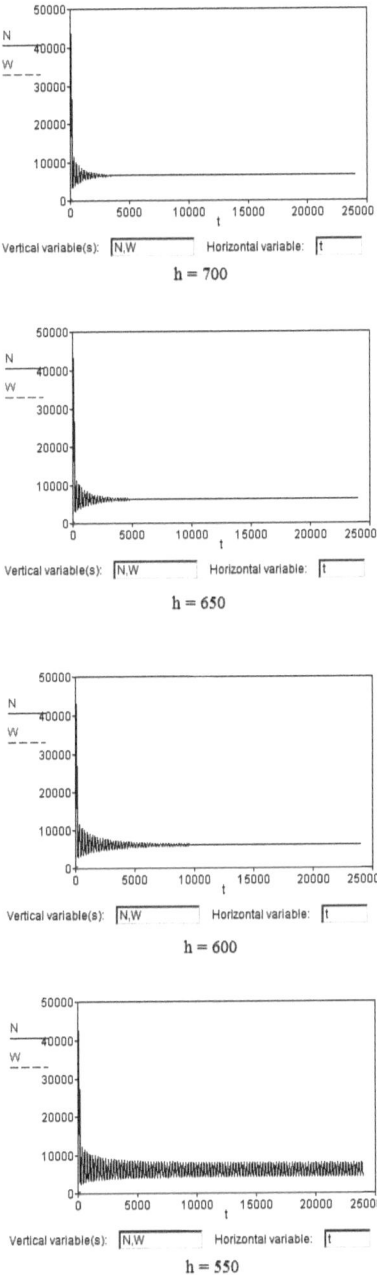

Figure 15.2: A Hopf bifurcation: the appearance of a limit cycle as *h* decreases.

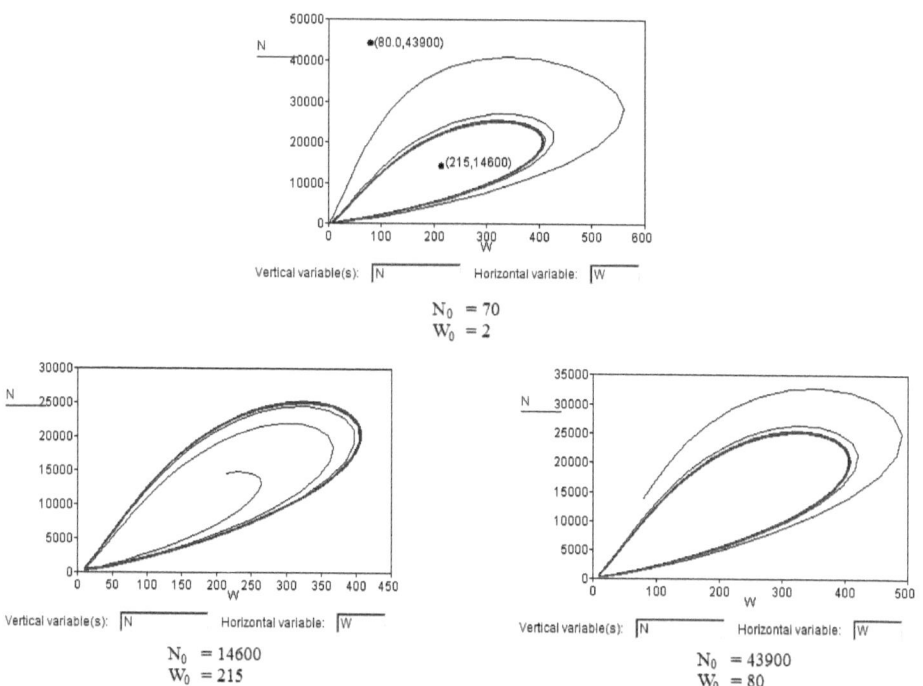

Figure 15.3: Phase portraits of the same system with different starting values, all of which tend toward the same limit cycle.

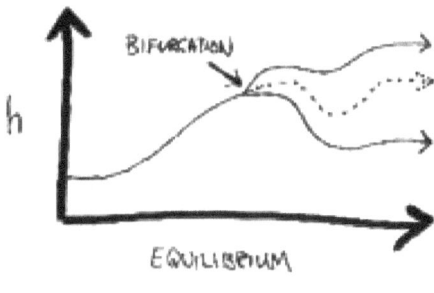

Figure 15.4:

Figure 15.4: Casual sketch of a bifurcation diagram. The amplitude of the Hopf bifurcation is given by the two solid curves denoting high and low points of N (P could also be used).

in the stability of a fixed point is typical of a Hopf bifurcation. Sometimes this behavior is depicted in the literature by graphing the location of the equilibrium (for N say, or for both N and W in separate pictures) versus the value of h. The line is solid until the point at which the equilibrium becomes unstable, after which it is dashed. The bifurcation point is usually labeled somehow. After the bifurcation occurs two solid lines indicate the amplitude of the limit cycle, the lower line giving the value of the trough and the higher one the peak. All of these values can be read off of graphs numerically generated by computer, although it is often easy to compute equilibrium values. Figure 15.4 shows such a sketch for these runs, with a smooth curve connecting the values.

Other graphs people use to describe this phenomenon plot the period of the oscillation against h, or sometimes vary more than one parameter at once to designate a region in a parameter space (such as the (h, j) plane) where a stable limit cycle exists. One very important aspect of any such study is varying the starting values of N and W to see if the behavior is independent of these, as it is in this model.

One thing we learn from researching the behavior of the system with respect to its parameters is how much we can trust our conclusions in the face of uncertain knowledge of some or all parameters. In this case the least trustworthy parameter, h, can have a big effect on the qualitative behavior of the system. On the other hand, the value chosen for h (200) is not close to the Hopf bifurcation. We could double this value without changing the long-term behavior of a limit cycle. So if we knew, for example, that h was fairly certain to be between 100 and 400, with some justification that it was near 200, we would be reasonably safe in concluding that the natural system ought to display a stable limit cycle.

The presence of a limit cycle is an important feature of this and many other systems. In the context of controlling water hyacinth, it tells us that when levels of hyacinth are particularly low or high they are not going to stay that way. It also tells us that decreasing the quantity of water hyacinth in such a way that it pushes the values of N and W outside the limit cycle will result in a subsequent value of N that is larger than that predicted by the limit cycle, until the system returns to its stable state. It also tells us that such cycles are an intrinsic part of the predator-prey dynamic and not necessarily due to outside forces such as weather, although those would certainly affect the system somehow.

Generally this model predicts that the weevil will be quite effective. Without weevils present, the limiting value of water hyacinth is 70,000 grams per square meter, which is observed in nature. With the weevil present and even in the presence of a limiting cycle, the amount of hyacinth drops dramatically. In nature this also appears to have happened since the introduction of the weevil to Lake Victoria. It would be interesting to know if local data support the existence of the stable limit cycle predicted by Wilson's model.

Chapter 16

The Uncommon Cold

Since antiquity, the ability of mathematics to predict natural phenomena has been the subject of a certain amount of wonder. Why should the constructions of the human mind mirror the natural world so well? Galileo claimed that the language of the natural world is mathematics, and all natural laws are written in that language. In saying this, Galileo declared himself a Platonist, one who believes that mathematical objects, (such as perfect circles, squares, and the logistic equation), have a reality independent of the mind that perceives them. If Galileo is correct, and the mind only perceives mathematical structure, then what is the organ of perception? No one has an answer to this question. If, on the other hand, the mind is inventing mathematics, as it does poetry or music, then why is this invention such an accurate mirror of natural phenomena? No one has an answer to this question either.

One of the great powers of mathematics is its capacity for describing a multiplicity of situations with the same expression. Consider the logistic equation, as in Figure 16.1.

$$\frac{dP}{dt} = kP(C - P)$$

We have seen this earlier in the context of limited growth. P is the population of a species, and $C - P$ is the room left for growth. C is the capacity of the system to support the species whose population is P. Yet, this equation is not limited to one interpretation.

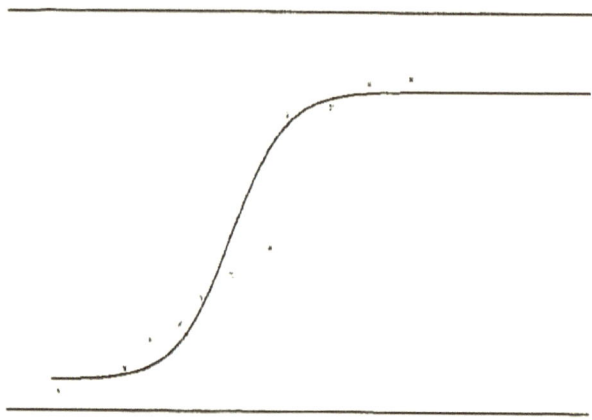

Figure 16.1: Typical results from our experiment.

Let us do an experiment. Suppose each person in a room of 25 or so people is given a card and a die. At random one person is designated "infected" with a contagious disease. This person marks his or her card with an I. Then people walk around the room at random, shaking hands. At every shake, a die is thrown. If the number is three or below, no transmission occurs. If the number is 4 or above, both parties compare cards to see if one is infected. If so, the other becomes so. Each person should record their trials, marking the point of infection with an I.

After forty or so trials, graph the results as trials versus number infected. The first trial has only one person infected. After this, the number of infected people grows. Of course, there can never be more than the total number of people in class who are infected. So after a while, the histogram levels out at the 25 person mark. Usually the graph that appears on the board after this experiment has a certain resemblance to the S-curve associated to the logistic equation, as in Figure 16.1.

Why should this be so? The key lies in viewing the situation from the point of view of the organism causing a disease. A human pathogen is a creature requiring a particular habitat like any other creature.

In this special case, the population of humans constitutes the habitat. Transmission occurs when an infected individual contacts an uninfected one. If such an occurrence is random, then its probability of happening is the product of finding an infected person and an uninfected person in the same location. The probability of independent events occurring simultaneously is the product of the probabilities of each happening separately. So, the P in the logistic equation is proportional to the probability of finding an infected individual, whereas the $(K - P)$ is proportional to the probability of finding an uninfected individual. $P(K - P)$ is the chance of the two coming into contact at a given time.

From the point of view of the disease organism, its ability to grow is proportional both to its population and to the space left for growth. A reasonable equation governing its behavior is therefore the logistic equation. The logistic equation is as well known in epidemiology as it is in ecology.

This model is based on some assumptions. First of all, the disease must be transmitted from human to human directly. Many diseases, such as the common cold, are transmitted directly. In fact, a cold can be transmitted by hand to hand contact, such as is suggested in the above experiment. If one person in the room really does have a cold, the exercise described above is no mere simulation. In fact, almost every person in the room will have the germs for the cold on their hands. If you do this experiment with friends or in a class, be sure to wash your hands afterwards!

Another assumption that is made by our simulation is that of randomness. Every person in the room is approximately equally likely to contact any other person. This is true of some situations, such as populations of students eating in dining halls. But it is not true of other situations, such as the population of extremely poor people versus the population of well off people. Nor is it true of sexually transmitted diseases. In modeling these diseases one might assume that individuals are quite promiscuous within a specific group, while contact between groups would be less probable (although not zero!).

Yet another assumption about the disease in question is that an infected person remains infected (and thus able to infect others) forever. This is not the case with most diseases. Usually the body's own immune system defeats the virus or bacteria after a while. In this situation, one would have to take into account three populations of people: those who have not yet had the disease, those who are infected and contagious, and those who have recovered and can no longer either get the disease or transmit it to others. The common cold, the flu, and measles all fit this scenario better than the simpler one, although transmission rates are so quick that the portion of those who have recovered might not matter much. Yet, a health care center might want to know how many people are likely to be actively sick at one time.

Finally, we have made the unspoken assumption that we are dealing with a single disease. Some diseases, such as the common cold, or HIV, or malaria, mutate rapidly. Mutation allows them to circumvent the immune system, returning for repeat infections by repeated transmission or, as in the case of AIDS, thwarting the body's own internal immune process by mutating inside the host.

Each of these modifications of our assumptions would result in a different model for disease transmission. In each case, known patterns for the disease in question should be checked, if possible, against predictions given by the model. The more complicated the model becomes, the more kinds of qualitative output it might generate. Thus, as the equations grow in complexity, it becomes imperative that the constants be identified correctly, both as to their meaning and their true numerical values.

As an example, the reader is invited to consider the problem of administering measles vaccine in a large university where an outbreak is imminent. In this situation (which actually occurred), most students were not vaccinated against a particular strain of measles. Transmission of measles occurs from person to person, as with the common cold. Several cases were diagnosed, all within a particular dormitory and dining complex. The students living in that dorm were immediately quarantined, and all available vaccine within in the state was sought in anticipation of a major measles epidemic.

The main problem facing student health services was one of limited resources. Measles vaccine degrades rapidly, so the entire supply available was only a fraction of the student population at the university. In a lucid moment the doctors approached the mathematics department. The initial inclination of the doctors was to vaccinate everyone in the quarantined area. Soon they realized that this would waste a lot of precious vaccine because many of those students were already exposed to the disease. How could they maximize the benefit of limited resources? Graduate students assigned to the problem produced a box model including susceptible, contagious, and immune (or recovered) students in each of the main dormitory-eating areas in the university. They wrote down equations governing the system and ran them on the computer. They modeled the effects of administering vaccine in various ways around campus. The reader is invited to invent such a model on their own, to see if it sheds light on the problem. Theirs was a version of the classic SIR model found in all the literature on epidemiology.

Ultimately, the course that the mathematicians recommended was not followed. No one doubted that it would maximize the healthy population of the university and minimizes cases of measles. The problem was that it was not intuitive, therefore it could not be explained, (by mathematically weak college administrators), to concerned parents whose children had been denied the vaccine. In fact, the doctors were forced to do the very least efficient thing with the vaccine, thereby wasting much of it.

The common cold is an example of a disease with a simple person-to-person transmission pattern. Many diseases rely on insects or other animals as vectors of the disease or reservoirs of disease. Here are a few examples of diseases that require more complex models.

Bubonic Plague

Historically the Bubonic Plague is famous for a European epidemic in the middle ages, but the Bubonic Plague is still present in the modern world. The plague is spread with mammals and fleas as vectors. For information on how it is spread see Boyd & Kemmerer (1921), Dunn (1923), Salkeld *et al* (2007), and Ward (1910). For a mathematical model of the Plague, see Keeling & Gilligan (2000), and for general disease modeling, Keeling (2001).

West Nile Virus

West Nile Virus is another disease which can be modeled in

respect to hosts and vectors. For a general overview of the disease and models, see Wonham *et al* (2006). For another model, also look at Bowman *et al* (2005).

Cholera

Cholera is another disease, which can be spread through water, and can be modeled mathematically. It is present in the Lake Victoria region. For information on the disease, see Longini *et al* (2002). For mathematical models of cholera, and its spread, see Codeco (2001), Vital *et al* (2007), and Huq *et al* (2005).

African Cassava Mosaic Virus

The African Cassava Mosaic virus affects the Cassava plant, which is a major source of food in Africa. It is spread by the whitefly. Modeling plant diseases present additional challenges and simplifications in modeling. For a basic overview of the African Cassava Mosaic Virus and work that has been done so far see Thresh & Cooter (2005). For some modeling examples of the spread of the disease, see Fargette & Vie, (1994), Holt *et al* (1997), and Van den Bosch *et al* (2007).

For your consideration

Question 1:

Is this experiment a good probabilistic model for the logistic equation? Does it accurately reflect the hypotheses that lead to the equation? Could you improve upon it? How?

Suppose we change the rule for "transmission" from rolling 4 or above to rolling 5 or above. How does this change the model, both in terms of what that rule represents, and also in terms of the expected outcome? Changing what term in the logistic equation would correspond to changing the rule for the die?

The match between data for the experiment and the logistic curve itself is not perfect. Real data never matches a curve perfectly. And yet, the curve is a good predictor for this phenomenon. Suppose you have some epidemiological data (like the simulation data in class) and you want to fit a logistic curve to it, in order to determine the transmission rate. In other words, you know P and C in the following equation:

$$\frac{dP}{dt} = kP(C - P)$$

But you also want to know K. How might you go about fitting the data to a particular logistic curve?

Question 2:

During an outbreak of a contagious disease, epidemiologists often break the population into three hypothetical groups: susceptible, infected, and recovered (immune). Make a box model and set of equations that describe this situation. Three boxes would be required. What would the equations look like? What happens when you implement this model? Experiment with your model to see how the parameters affect the outcome. What do the various constants represent? What happens when rates of transmission or recovery change? Does the output always look the same, qualitatively? These models (known as SIR models) are all over the literature, and you can research different ways in which they are used.

If you were advising a clinic about this epidemic, what would you tell them? They would want to know what the traffic would be like, how much medicine to keep on hand, whether to hire extra nurses and when to do so. Use your model to advise them.

The manager of a health care provider might want to know whether encouraging people to wash their hands (thus lowering the transmission rate) would be enough to lower the peak population of sick individuals, thus spreading the burden of health care over a longer time interval. Would it, based on the model developed in this problem?

An easy addition to the SIR model is to add a box for quarantined individuals. Quarantine has been used for centuries to control the spread of disease. Some sources say the first example of quarantine was an island near Dubrovnik off the coast of Croatia, where people were sent who showed symptoms of plague. Dubrovnik also had a bunkhouse just outside the city where all travelers were quarantined for a period to see if they had any disease before being allowed into the city. What does the rate of quarantine do to the duration of an epidemic?

Question 3:

Suppose we have two populations that rarely come into contact. What would be the box model for transmission of a contagious disease in this case? What would the system of equations look like for your model? Use software to see what kind of pattern your model predicts. This situation is similar to what has happened with HIV in this country, as one sub-population after another is introduced to the disease. Formerly considered a disease of homosexual males only, HIV now infects every demographic group.

Question 4:

For each of the questions above, think of a variation on the simulation described in the chapter that is based on the same assumptions as those captured by the model you have invented. Run it and see if the data resembles that of the computer model. If not, analyze what went wrong.

Question 5:

Spacial factors might be important, as in the college dormitory measles problem. If the university in question had six large dorms and disease broke out in one of them, how would you deploy a vaccine? This calls for a box model and a set of equations. Suppose that there is only enough vaccine for one of the six dorms. Does it make sense to spread it around or vaccinate everybody in a single dorm? What if some dorms were less likely to contact the infected one, so transmission rates vary between locations in some way. Does it make sense to vaccinate the one most likely to have had contact with the infected students? See what your own model says about these questions.

Chapter 17

Swamp fever

It is hard to know how to begin the story of a tragedy. Perhaps for western readers it makes sense to start with a familiar tale, that of Pharaoh and the Jews. According to biblical tradition, Pharaoh attempted to murder all the firstborn sons of Jewish families living under his rule in Egypt. This act was a small dose of murderous genocide, enough to bring on the plagues and result in the migration of the entire Jewish population to Israel. Even by today's standards of science fiction tall tales and horrifying ethnic "cleansing", Pharaoh's attempt remains a benchmark of horror. Jewish liberation is commemorated every year for good reason.

Malaria kills 3,000 children under the age of five, every single day. These children live in tropical and subtropical areas such as the region around Lake Victoria. Africa hosts ninety percent of all malaria cases. In the region around the lake approximately one child out of twenty will die of malaria before reaching the age of five. The complete hopeless misery of the situation is difficult to communicate, but various sources attempt to do so:

3,000 children per day

One child every 30 seconds

Seven jumbo jets full of children, crashing every day

One child in twenty

One out of four deaths of children under 5

One out of ten infant deaths

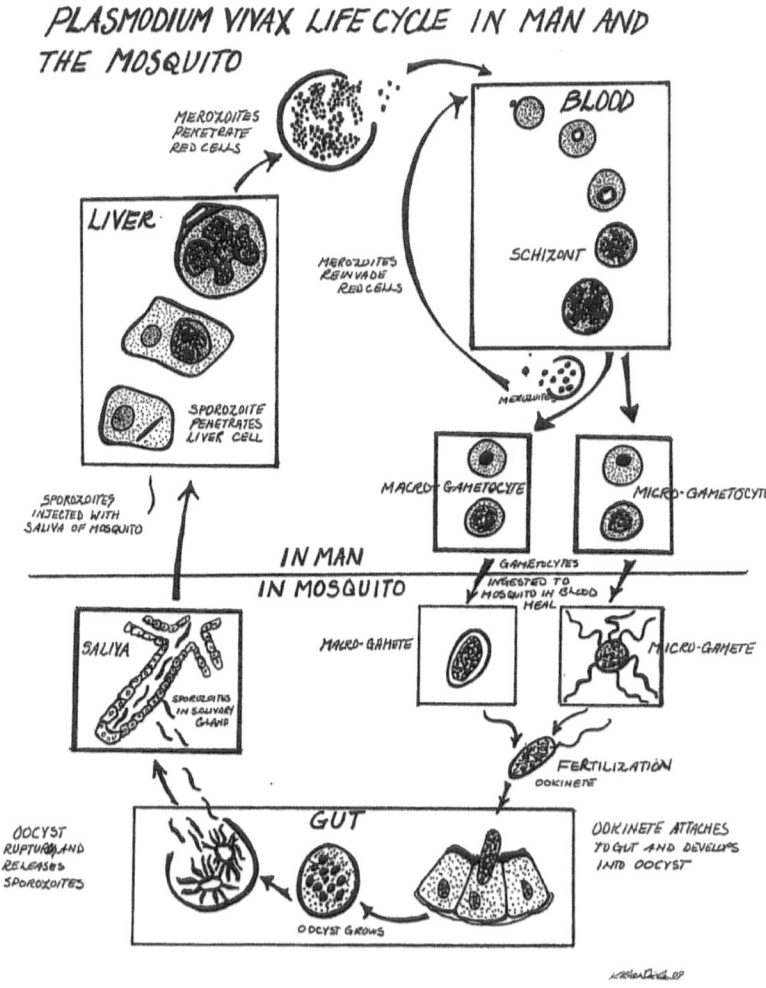

Figure 17.1: Plasmodium life cycle. Art by Kim Rheinlander

Pick your favorite. The impact on the adult population is also documented, with estimates of 270 million people infected by the disease every year. 1.5 to 2.7 million die from it each year, including the children. Recovery takes several weeks. In some areas, malaria sufferers occupy a third of all hospital beds. Families sickened by malaria can only farm forty percent of the land they could cultivate if healthy. A twenty percent tax on Gross National Product by a colonizing country would be an insufferable situation leading to revolt, yet the gross national product of a nation suffering from malaria is reduced by about twenty percent according to some estimates.

Clearly malaria is a disaster on personal, family and even national levels. In the 1950's, the World Health Organization initiated the "Global Eradication of Malaria Program", which operated until 1969. By 1967, the organization had shifted its perspective to one of "control" of the disease. Eradication seemed impossible. The search for drugs and vaccines has had limited success, yielding the familiar quinine and a few other options. Part of the difficulty lies in the population affected. Poor people in poor countries cannot afford expensive medicine; thus pharmacology companies find malaria a poor investment. Research goes on at non-profit institutions such as universities and hospitals, yet nothing approaching a "cure" is at hand. A good researcher knows that, whether the goal is to cure or control a disease, understanding it well is the only viable first step. The literature is full of mathematical models of the life cycle of this amazingly complex disease.

Africa is the birthplace of humankind. Because of this we necessarily recognize that Africa is also the birthplace of many of mankind's various diseases. Humans and their diseases co-evolved in Africa, dancing together in patterns made more intricate as the centuries passed. There are many types of malaria, of which only *Plasmodium falcifarum* is regularly fatal. A good parasite should not, for its own sake, kill its host. The fatal form of malaria is less well evolved than its less lethal cousins.

The organism that does the damage in humans is a protozoa called Plasmodium that has a complex life cycle involving both humans and mosquitoes. The "deadly fever" characteristic of the disease was recorded as early as early as the very earliest writings, about 6,000-5,500 B.C. The association with swamps was well documented, but the first sighting of the malaria parasite was not made until 1880. The role of the mosquito in the transmission of the disease was finally established by 1898.

During its life cycle the protozoa passes through numerous forms, both asexual and sexual. At its point of entrance to the human host from the mosquito, the parasite is in the form of a *sporozoite*, which enters human blood from the saliva of a hungry female mosquito. The *sporozoite* takes a free ride to the liver, where it penetrates a liver cell (*hepatocyte*). Inside the liver cell the *sporozoite* multiplies for 9 to 16 days, emerging in a form, the *merozoite*, that can infect red blood cells. In some species another form of the organism, the *hypnozoite*, remains behind in the liver and causes a delay of *merozoite* release, but not all species do this. The *merozoite* then infects red blood cells and multiplies within them until they burst, releasing the protozoa in three forms. The same form, the *merozoite* repeats the life cycle inside the human blood. Some of the *merozoites* will have changed to "sexual" forms, the *macrogametocyte* and *microgametocyte*. These forms do not re-infect the human host, but pass to a mosquito during feeding. Until the production of *gametocytes*, the human is not contagious.

A female mosquito feeding on a contagious human will suck up the *macro* and *microgametocytes*, passing them into its gut, where the *macrogametocytes* are fertilized and become *ookinetes*. The *microgametocyte* fuses to the cell of the *macrogametocyte*, and its nucleus enters the cytoplasm of the *macrogametocyte*, where fertilization occurs. Lodging in the wall of a cell in the mosquito's gut, the resulting *ookinete* becomes an *oocyst*. *Sporozoites* form inside the *oocyst*, and are released when it ruptures. From there they migrate to the salivary gland and are released when the mosquito bites its next human, thus completing the reproductive cycle. Unless the mosquito has *sporozoites* in its saliva, it is not contagious.

The function of the spleen is to remove dead red blood cells from the body. The spleen acts as a sieve. Healthy blood cells are flexible and can bend in two, passing easily through the sieve. Used up or diseased cells become hard and are caught by the spleen, dissolved and removed. The production of merozoites in red blood cells creates a heavy load for the spleen to remove. The characteristic temperature spikes associated to malaria are associated with this build up of infected cells, as is the presence of an enlarged spleen.

Intervention during the human portion of the *Plasmodium* cycle has been through drugs, primarily quinine and associated synthetics, which kill the protozoa inside the human. Unfortunately, these drugs do not prevent re-infection. Over time a human develops immunity to a particular form of malaria, but only if constant re-infection occurs. Thanks in part to the sexual portion of its life cycle, *Plasmodium* mutates rapidly to bypass the human immune system. Therefore, a person who has considerable resistance to malaria, and who leaves the infested area for a few years, returns as susceptible as an infant. This is the reason travelers suffer symptoms from malaria more severe than residents who are constantly exposed. It is unclear whether the disproportionately high death rate among children could be explained entirely by the fact that many infections are first infections in that population, or whether the physiology of children must be taken into account. The rapid rate of mutation also means that varieties resistant to the standard drugs are beginning to appear.

Each type of malaria has its own specific mosquito (*Anopheles*) host. Evidently mosquitoes have built up a better set of defenses against malaria than humans, as each mosquito species can only be infected by a single strain of malaria. In any case, the epidemiology of malaria has everything to do with the life cycle of its mosquito host. The mosquito lays eggs in stagnant water. These hatch into nymphs, which continue to live in water until they change into adult mosquitoes. Only the adults transmit malaria. Furthermore, the predators of the mosquito are different at each of the stages of the cycle. Fish eat the nymphs, but bats and birds eat the adults. Epidemiological intervention during the mosquito portion of the *Plasmodium* life cycle has an old history. The ancient technique of draining all swamps was replaced in the 1950's with coating marshes with a layer of paraffin or simply spraying the areas with DDT.

A chapter on a subject as complicated as malaria can only be a pointer to the vast literature on the subject. Medical libraries are full of articles about malaria, many of which contain mathematical models. The questions at the end of the chapter are not "math problems" in the usual sense, but are meant to guide a deep mathematical, environmental and medical research project done collaboratively by an entire class. Indeed, a single person cannot solve the problem of malaria. Some of the questions are suggested by claims made by the World Health Organization and others. As you study the literature, you will find other claims that can be tested by a model. Each claim represents a research question for you to consider if you wish.

The true value of mathematical modeling becomes apparent in a situation like this one, where the "right answer" cannot be known, and all "answers" to date have proven temporary at best. Mathematics is a crystal ball, but the answers it gives depend precisely on how we ask it a question. We offer hypotheses about how a system works and mathematics tells us what must follow from those hypotheses. Whether we get useful information depends on our ability to think critically about both the equations and the output the computer yields. Some estimates for the number of malaria-infected individuals at any given time run to 500 million people. This number is a sizeable fraction of the entire world population, and represents a degree of misery that is hard to imagine. If we are experiencing global warming, then the habitat for the malaria mosquito is likely to increase, leading to the worsening of an already bad situation.

It is worth comparing the problem of malaria with that of AIDS. There is a point of similarity, which is the rapid mutation rate of both HIV and Plasmodium. This mutation rate is a major complication in the epidemiology of both of these diseases. However, in the case of AIDS, curing a single case of the disease is very difficult (and merely controlling the progress of a single infection has proven extremely expensive), while a simple change of human behavior (using condoms, for example) can halt the transmission of the disease completely. For malaria, the situation is reversed. Curing a single case is easy and inexpensive-just take your quinine or tetracycline. But due to the complexity of the cycle of transmission and the mutation rate, there is no easy intervention that will halt the transmission of the disease completely. So, in addition to the epidemiology, the economics and politics of the two diseases are quite different.

No references are provided for the literature on modeling malaria here. There are many, and part of the researchers job is to find these. The point has come in this text to force the reader into just such an exercise.

For your consideration

Question 1:

A single individual would find the problem presented in this chapter daunting. But a team of ten or twenty people could frame a tractable approach. The best way to begin is with a list. What are the factors involved in understanding malaria? Developing this list as a team is the best way to start studying this disease.

Once you have a list, it would be advisable to separate it into smaller components. For example, the life cycle of the mosquito will surely be on your list. This is something that can be studied separately from the *Plasmodium* cycle, although the effect of the disease on the mosquito population might need to be taken into account. Take your long list and create three to five subsystems that can be studied effectively. Divide yourselves into teams, one for each of these problems.

Question 2:

For each problem it will now be necessary to develop a model. The model must be based on what we know about malaria, humans, and mosquitoes. Any constants needed (such as incubation periods, transmission rates, etc.) must be researched. You will find them in the literature, because malaria has been fairly well studied. You will also find many mathematical models to consider. Some will be justifiable and others will not. Developing your own model using the best features of those in the literature is the best way to get an answer you really believe.

Before using your model to predict anything, it is necessary to test it against what is already known. You should make a list of things that your model ought to predict if it is a good model. For example, a model of the human physiology of malaria ought to predict regular temperature spikes that slowly decrease in intensity. These spikes are believed to correspond roughly to the amount of parasite in the blood (or liver). The time between spikes is closely associated with particular forms of *Plasmodium*. Various constants in your equations should be adjusted to match the known data. Any epidemiological model needs to take into account the growing resistance to the disease, as well as the effects of leaving and returning without immunity. There might be other known features of malaria that your model should predict.

Also, you should know the range of possible behaviors for your models. Are there repeating cycles, as with the Lotka-Volterra model? How many different long-term behaviors can your model have? Do small changes in initial values create large differences in behavior? How well can you trust the constants you have found? Would a small change in one of these make a difference in the kind of long-term behavior your model predicts? These are standard mathematical questions one wants to know about for any model.

Question 3:

Once you have some models, you can test various hypotheses. The World Health Organization claims that the ill effects of malaria could be greatly reduced if everyone slept under mosquito nets at night. Is this likely? The use of DDT on swamps has been largely discontinued. Is there a reason, based on your model, that it might backfire in the long run? Some have proposed developing disease resistant mosquitoes and introducing them into the swamps, hoping they would compete with the diseased varieties, leading to the eventual elimination of those. Would this work? The World Health Organization also claims that, with enough money, malaria could be eradicated. What exactly does this mean? What would it take to interrupt the life cycle of the disease so completely that it is actually eradicated?

The word "eradicated" requires some expansion. It is very unlikely anybody can kill all the mosquitoes in a given region, unless all swampy areas are drained (as the Romans did). What is actually happening is that populations are reduced to the point where transmission rates of malaria are so low that the disease cannot persist. If malaria followed a simple logistic model, this phenomenon would be impossible without total elimination of the disease. But more complex models could display an Allee effect, which is a highly desired phenomenon for diseases because it implies that reducing the disease below some threshold is enough to result in its disappearance. At this writing the author has not seen any epidemiological models for malaria that are known to have an Allee effect.

Question 4:

Finally, time permitting, it would be worth trying to make two or more of your models work together. For example, if you put mosquito nets over everyone, you lower the transmission rate from mosquito to human. This is an epidemiological question. Presumably you also lower the transmission rate from human to mosquito. If a competing strain of mosquitoes, immune to malaria, is introduced into the region, how will the mosquito nets affect their ability to compete? (This is not a hypothetical suggestion–such a species has been identified and this path has been proposed as a solution.) The ecological model for mosquitoes must have a conversation with the epidemiological model to settle on an answer. Malaria is a large, complex system. It is certainly possible that intervening in two parts of it in an uncoordinated way might be worse than doing nothing.

Question 5:

http://www.cdc.gov/epo/mmwr/preview/mmwrhtml/00000793.htm
September 12, 1986 / 35(36);567-8,573
Outbreak of Malaria Imported from Kenya
Malaria remains an important problem for U.S. travelers to areas in Africa with chloroquine-resistant Plasmodium falciparum, as illustrated by the following report:
On June 26, 1986, a 29-year-old resident of Louisiana became ill en route to the United States from Nairobi, Kenya. He was admitted to a hospital in Shreveport, Louisiana, with a diagnosis of P. falciparum malaria on June 28. By June 30, seven additional persons with malaria-like symptoms were admitted to three Shreveport hospitals. All patients received medical care within 2 days of the onset

of illness; P. falciparum parasites were identified in each patient's blood smears within 2 days after being seen by a physician. All were treated promptly and successfully with quinine and tetracycline.

Based on this news item one might well ask: Why don't we just give Kenya a large donation of quinine and tetracycline?

One thing that models are very good for is predicting what *will not* work. Suppose you are an epidemiologist for the World Health Organization. A large shipment of quinine is donated to the "war on malaria". Imagine several options for deploying this medicine. Develop an epidemiological model to predict which one will result in the most lives saved. Keep track, as you go, of suggestions that definitely will be of no use. As an amusing entertainment you might want to look up historical documents that tell how medicine was actually deployed under similar circumstances. You may find that the history of good will contains a lot of well-meaning actions doomed to failure. Perhaps you will conclude that saving three thousand young lives a day may well rest on the courageous and ruthless commitment to the art of seeing what is really there.

Afterward

One striking aspect of writing a book like this is coming to terms with vast world of topics one is forced to omit. We could have had some great chapters on HIV, plague, cholera, and many other diseases. We could have pushed the mathematics much farther, giving the reader an introduction to tools such as the Hartman-Grobman theorem that allows us to compute stability of equilibrium points, explorations of the dynamics of chaos, or spacial models relying on nonlinear systems of partial differential equations. We could have looked at geographic regions other than the area around Lake Victoria. In my defense I can only say that I have included topics in mathematics and biology that have worked well with my students, although their research usually extends beyond the topics I chose for the text.

Mathematical biology, and in particular the kind of modeling in this text, is a natural aspect of many fields of research, including applied math, public health, bioengineering, biostatistics, pharmacology and medicine in general, ecological modeling, environmental studies and even economics or earth science. Anyone desiring advanced material in this area, or interested in pursuing a related career, can search across a variety of departments and programs for more courses and degree programs.

Most of my students are preparing for careers in medicine and many of them will become researchers at the forefront of their fields. Medical research is subject to the economic stresses that affect us all. Because of the demographics of the population in the States and because of inequities of income, right now the research money is heavily weighted toward what I jokingly refer to as "geezer medicine".

("Geezer" is a slang word for a senior person, and not such a nice word, really.) As a future geezer myself I expect to benefit from the hard work of my students and I would love to walk into a doctor's office and find one of them there. (I like to think they would be pleased too.)

But I must recognize that the majority of suffering in this world is not due to geezer disease. It is, as we all should know, almost a privilege to live long enough to contract cancer and heart disease. Malaria, cholera, HIV, and other infectious diseases, take a far greater toll on the people of this world than the extensively funded problems of geezer medicine. So the topics chosen for this text are also a political statement. I always hope that my students will consider taking a path that resists the inevitable push of funding and will make a conscious choice to improve the lot of those who cannot necessarily pay full price.

Notes to the Instructor

The great thing about teaching mathematics in the context of medicine and biology is that the context itself solves the problem of motivating students. Tropical vector-borne diseases in particular offer many opportunities to bring students to the point of fascinated disgust. I personally think this is a great place to be. I love the fact that plague-ridden fleas backwash when they bite you, and that the parasitic worms responsible for schistosomiasis mate for life inside your body.

Context also offers a tool for approaching the mathematics. When teaching force diagrams in physics it helps to put yourself in the position of the object on which forces act so you can appreciate which directions you are being pulled and pushed. The same principle works well in this subject. To make a good algebraic term to describe a particular rate, it helps to put yourself in the position of the organism or body part that is responsible for that rate. Understanding how the organism reacts to its situation is the basis for the algebraic term in the differential equation. If you can't write that term down, then you really don't understand the organism. Systems also function as a tool for refining our understanding of basic biology. If we try to teach mathematical biology by giving the students the equations that "are known" to describe a phenomenon, we deprive them of the most important aspect of the subject. Mathematics is fundamentally a sense-making activity. This is why humans do it.

Goals My goals for students in my course are two-fold. First and foremost I want them to be able to read and think critically about a body of research literature that includes systems of ordinary differential equations in the service of biological problems. I can tell whether they are able to do this by the kinds of papers they write for me. Secondarily I have a more important but slightly fuzzier goal of giving students the opportunity to engage in real research, even in the very short time frame of a nine week quarter. My students write three papers (some done as group projects) during this short time, resulting in a fair amount of exposure to the literature and three abbreviated forays into research. The text is not written in the same style as research literature. If all of my students could handle that literature when they walked into the class, there would be no need for this text at all. The text is intended as an informal, friendly guide into the real stuff. So I have to make sure my students understand that their papers should emulate the literature and not the text.

Pedagogy I would never tell you how you should teach. But maybe it would be useful if I tell you how I teach, since this textbook serves a course that may not yet exist for you.

I want my students to be producers of knowledge and makers of models, not just consumers of these. In a short period of time I want them to invent, explore, analyze and critique more models than at first appears possible. You might say that each chapter in the text is a case study. I let them do most of the work during our two-hour class periods. I am assisted in this by the natural tendency of students not to read ahead, a tendency I prefer to leverage rather than fight. To the instructor it feels like things are going ever so slowly. There is a huge temptation to take the problem away from the students and do it for them. Fry it (or them) instead of stewing. But paradoxically, we can only go fast by going slow.

I try to lecture only briefly to set up a question, which students may then use considerable time to address. It works best for the students to begin with drawing a box model that is a pictorial representation of a system. As a class we can compare different pictures and usually come to a unanimous agreement as to which one best represents the system under discussion, if the system is simple enough. There are a few box models in the text but not many, because I would prefer to have the class generate them.

I might summarize such a discussion by saying (as an example)

"I see that there are basically two box models you like. One of them has boxes for susceptible, infected and recovered people, but the other also includes a box for exposed individuals who have not yet developed symptoms. Both of these models are valid and you will find both in the literature, depending on what the disease is. Basically, if the time is short until symptoms appear then models tend to ignore the 'exposed' box. But if the time is long, or if human behavior is very different in those two groups, then the 'exposed' box could play an important role in the model. Can you think of an example where the 'exposed' individuals who didn't know they were sick would play a key role? Yes, HIV, exactly. If you were modeling the epidemiology of HIV you would include this box because of the long latent period and the higher rate of transmission during it, as people probably wouldn't change their behavior until they knew they were infected. (So use condoms.) HIV models sometimes have yet another box to distinguish between infected individuals and those with full-blown cases of AIDS."

After we agree on a box model, the students (usually working in groups) suggest what form each term in an equation ought to take. I always try to make sure the whole class looks at every proposed model, because you learn a lot by figuring out what is wrong with models (and generally we suspect that they can't all be right). Looking at material that is very probably incorrect is an educational experience we usually deny our students but it is the key to developing critical skills. If everything you read is always right, it is hard to learn to question. Again, a discussion ensues that relates each term to what assumptions are behind it and to our basic knowledge of what the system must do to be believable. For example, I might gently point out a flaw in a growth term by saying something like "This term is positive even when the population is zero. So the population will grow even when none are present to start. So this model might be very good when the population is large but I guess I wouldn't trust this model near low levels of population, unless we all believe in spontaneous generation. That theory was discredited, right? I mean, you guys know more biology than I do."

Once we have a system we believe, the next step is to put it on the computer and see what it does. We use a simple piece of software that can integrate up to twenty O.D.E.'s and display a limited selection of outputs. People bring laptops to class. If your students don't all have laptops, then it would be a good strategy to conduct class in a computer lab of some sort. Usually computer explorations require students to answer specific questions about a model or carry out some procedure to that end. Early in the course we might do something like this but the real question is which aspects of computer output are relevant to the research. Another way to put this is: Which features of the output are trying to tell you something and what do they say? Once we have a model working, we want to ask it questions. Here is where ideas about equilibrium, stability, sensitivity and bifurcation can be explored fruitfully.

One reason this course works is that a lot of class time is also devoted to working directly on the three research projects that students choose. Each project emphasizes one of the three parts of the research process I just described, although all the components are present. Intermediate tasks of generating a research proposal, a bibliography, and computer output, are useful in intercepting crises before they get out of hand. A few words of direction early on are worth hours of pointless struggle later. I am in favor of struggle, but not pointless struggle.

The Literature The literature is a moving target. I have a couple of papers I like us to read together as a class but when the students turn in their research papers I expect to learn about resources that weren't there last year. Electronic services have made research projects much more possible for undergraduates than ever. I usually ask a librarian to give an overview of library search engines and services as the students begin work on the second paper.

The perfect instructor would be omniscient, but few of us are. When I started teaching this course all paper topics were about pharmacokinetics, ecology of Lake Victoria, and malaria. One of each, in that order. The range of topics has grown and with it my familiarity with the literature. Now we have papers on crocodile reproduction, SARS, cholera, and tumor modeling. I never know what my students will want to do. But because the range grew slowly I have been able to cope with it. I would suggest a similar strategy for anyone just starting a course like this one.

Resources The Center for Mathematics and Quantitative Education at Dartmouth (MQED) maintains an online resource collection for undergraduate teaching. The math and biology section can be found at:

http://www.math.dartmouth.edu/ matc/eBookshelf/biology/index.html

Here you will find a link to the free online version of The Big Green Ordinary Differential Equation Machine. A better version is available from the developer at

http://www.bpreid.com

which allows the user to add data points and compute the error between a model and data. Many other resources on the MQED site could be useful when teaching material from this text.

All discussions of disease profit from an initial review of relevant material on the World Health Organization web site. I have not referenced the site directly in the text because I would have to do so for every single disease. It is usually the right place to start.

Students as Researchers This course has shown me that students in regular courses can be a valuable addition to a research program. In addition to providing a literature review (of whatever they choose to review), they can sometimes have very creative suggestions for new directions to pursue and, in some cases, use their papers to sort out technical details of a model such as sensitivity to parameters.

Costs This text is published independently for three reasons. First, no editor was allowed to interfere. This was an important factor in keeping the text short and therefore inexpensive. Second, self-publishing results in a textbook that is priced like a regular bookstore hardback or paperback. Textbook prices have gotten way out of hand and this is one way to keep them down. Third, I own the copyright. This has two implications. One is that I can revise at will. The second is that I can give permission to reproduce the book. If you teach in a part of the world where the cost of a twenty dollar text is prohibitive (more than a few hours of wage, for example), please contact me. I can give you permission to print out a fixed number of copies. We should never let cost prevent learning, any more than we should let cost prevent health.

Bibliography

[1] M. Albonico, H. Allen, L. Chitsulo, D. Engels, A.F. Gabrielli, and L. Savioli, *Controlling soil-transmitted helminthiasis in pre-school-age children through preventive chemotherapy*, PLoS Neglected Tropical Diseases **2** (2008), no. 3.

[2] T.P. Albright, TG Moorhouse, and TJ McNabb, *The rise and fall of water hyacinth in Lake Victoria and the Kagera River Basin, 1989-2001*, Journal of Aquatic Plant Management **42** (2004), 73–84.

[3] A.J. Atkinson, C.E. Daniels, R.L. Dedrick, C.V. Grudzinskas, and S.P. Markey, *Principles of clinical pharmacology*, Academic Press San Diego:, 2001.

[4] V.Y. Belizario Jr, M.L.E. Amarillo, R.M. Martinez, A.O. Mallari, and C.M.C. Tai, *Efficacy and safety of 40 mg/kg and 60 mg/kg single doses of praziquantel in the treatment of schistosomiasis*, Journal of Pediatric Infectious Diseases **3** (2008), no. 1, 27–34.

[5] A. Birkett, *The impact of giraffe, rhino and elephant on the habitat of a black rhino sanctuary in Kenya*, African Journal of Ecology **40** (2002), no. 3, 276–282.

[6] A. Birkett and B. Stevens-Wood, *Effect of low rainfall and browsing by large herbivores on an enclosed savannah habitat in Kenya*, African Journal of Ecology **43** (2005), no. 2, 123–130.

[7] DWA Bourne, *First Course in Pharmacokinetics & Biopharmaceutics*, Disponible en Internet en la siguiente dirección: http://gaps. cpb. ouhsc. edu/gaps/pkbio/pkbio. html.

[8] K.C. Bowles, S.C. Apte, W.A. Maher, M. Kawei, and R. Smith, *Bioaccumulation and biomagnification of mercury in Lake Murray, Papua New Guinea*, Canadian Journal of Fisheries and Aquatic Sciences **58** (2001), no. 5, 888–897.

[9] C. Bowman, A.B. Gumel, P. Van den Driessche, J. Wu, and H. Zhu, *A mathematical model for assessing control strategies against West Nile virus*, Bulletin of mathematical biology **67** (2005), no. 5, 1107–1133.

[10] M.F. Boyd and TW Kemmerer, *Experience with bubonic plague (human and rodent) in Galveston, 1920*, Public Health Reports (1896-1970) (1921), 1754–1764.

[11] N. Castro, H. Jung, R. Medina, D. Gonzalez-Esquivel, M. Lopez, and J. Sotelo, *Interaction between grapefruit juice and praziquantel in humans*, 2002, pp. 1614–1616.

[12] N. Castro, R. Medina, J. Sotelo, and H. Jung, *Bioavailability of praziquantel increases with concomitant administration of food*, 2000, pp. 2903–2904.

[13] T.D. Center, F.A. Dray, G.P. Jubinsky, and A.J. Leslie, *Waterhyacinth Weevils (Neochetina eichhorniaeandN. bruchi) Inhibit Waterhyacinth (Eichhornia crassipes) Colony Development*, Biological Control **15** (1999), no. 1, 39–50.

[14] T.D. Center, F.A. Dray, Jr, G.P. Jubinsky, and M.J. Grodowitz, *Biological control of water hyacinth under conditions of maintenance management: can herbicides and insects be integrated?*, Environmental Management **23** (1999), no. 2, 241–256.

[15] D. Cioli and L. Pica-Mattoccia, *Praziquantel*, Parasitology Research **90** (2003), 3–9.

[16] C.T. Codeço, *Endemic and epidemic dynamics of cholera: the role of the aquatic reservoir*, BMC infectious diseases **1** (2001), no. 1, 1.

[17] D.A. Cornwell, J. Zoltek Jr, C.D. Patrinely, T.S. Furman, and J.I. Kim, *Nutrient removal by water hyacinths*, Journal (Water Pollution Control Federation) (1977), 57–65.

[18] J.P.G.M. Cromsigt, J. Hearne, I.M.A. Heitk "onig, and H.H.T. Prins, *Using models in the management of Black rhino populations*, Ecological Modelling **149** (2002), no. 1-2, 203–211.

[19] D.L. DeAngelis, W.F. Loftus, J.C. Trexler, and R.E. Ulanowicz, *Modeling fish dynamics and effects of stress in a hydrologically pulsed ecosystem*, Journal of Aquatic Ecosystem Stress and Recovery **6** (1997), no. 1, 1–13.

[20] L.H. Dunn, *Fleas of Panama, Their Hosts, and Their Importance*, The American Journal of Tropical Medicine and Hygiene **1** (1923), no. 4, 335.

[21] H. Frohberg and S.M. Schulze, *Toxicological profile of praziquantel, a new drug against cestode and schistosome infections, as compared to some other schistosomicides.*, Arzneimittel-Forschung **31** (1981), no. 3a, 555.

[22] R.J. Geiderlv and T. Platt, *A mechanistic model of photoadaptation in microalgae*, MARINE ECOLOGY-PROGRESS SERIES **30** (1986), 85–92.

[23] M. Giorgi, V. Meucci, E. Vaccaro, G. Mengozzi, M. Giusiani, and G. Soldani, *Effects of liquid and freeze-dried grapefruit juice on the pharmacokinetics of praziquantel and its metabolite 4?-hydroxy praziquantel in beagle dogs*, Pharmacological Research **47** (2003), no. 1, 87–92.

[24] F. Guisse, K. Polman, FF Stelma, A. Mbaye, I. Talla, M. Niang, AM Deelder, O. Ndir, and B. Gryseels, *Therapeutic evaluation of two different dose regimens of praziquantel in a recent*

Schistosoma mansoni focus in Northern Senegal, The American journal of tropical medicine and hygiene **56** (1997), no. 5, 511.

[25] LL Gustafsson, B. Beerman, and YA Abdi, *Handbook of drugs for tropical parasitic infections.*, TAYLOR & FRANCIS, NEW YORK, NY(USA). 1987. (1987).

[26] T.A. Heard and S.L. Winterton, *Interactions between nutrient status and weevil herbivory in the biological control of water hyacinth*, Journal of Applied Ecology (2000), 117–127.

[27] M. Homeida, I.A. el Tom, S.M. Sulaiman, A.A. Daffalla, and J.L. Bennett, *Efficacy and tolerance of praziquantel in patients with Schistosoma mansoni infection and Symmers' fibrosis: a field study in the Sudan*, The American journal of tropical medicine and hygiene **38** (1988), no. 3, 496.

[28] A. Huq, R.B. Sack, A. Nizam, I.M. Longini, G.B. Nair, A. Ali, J.G. Morris, M.N.H. Khan, A.K. Siddique, M. Yunus, et al., *Critical factors influencing the occurrence of Vibrio cholerae in the environment of Bangladesh*, Applied and Environmental Microbiology **71** (2005), no. 8, 4645–4654.

[29] G. E. Hutchinson, *The paradox of the plankton*, The American Naturalist **95** (1961), no. 882, 137.

[30] M. Keeling, *The mathematics of diseases*, plus magazine **14** (2001).

[31] MJ Keeling and CA Gilligan, *Bubonic plague: a metapopulation model of a zoonosis*, Proceedings of the Royal Society of London, Series B: Biological Sciences **267** (2000), no. 1458, 2219–2230.

[32] J.V.D. Koppel and H.H.T. Prins, *The importance of herbivore interactions for the dynamics of African savanna woodlands: an hypothesis*, Journal of Tropical Ecology **14** (1998), no. 05, 565–576.

[33] M. Lisi and S. Totaro, *Algae-light interaction: Study of an approximated model and asymptotic analysis.*, Transport Theory & Statistical Physics **36** (20070601), no. 4-6, p323 – 349.

[34] I.M. Longini, Jr, M. Yunus, K. Zaman, AK Siddique, R.B. Sack, and A. Nizam, *Epidemic and endemic cholera trends over a 33-year period in Bangladesh*, The Journal of infectious diseases **186** (2002), no. 2, 246–251.

[35] RL Mackey, BR Page, KJ Duffy, and R. Slotow, *Modelling elephant population growth in small, fenced, South African reserves*, South African Journal of Wildlife Research **36** (2006), no. 1, 33–43.

[36] ME Mandour, H. El Turabi, MM Homeida, T. El Sadig, HM Ali, JL Bennett, WJ Leahey, and DW Harron, *Pharmacokinetics of praziquantel in healthy volunteers and patients with schistosomiasis.*, Transactions of the Royal Society of Tropical Medicine and Hygiene **84**, no. 3, 389.

[37] GG Marten, *Human Ecology: Basic Concepts for Sustainable Development.*, Earthscan Publications Ltd, 120 Pentonville Road London N 1 9 JN UK. 256 (2002), 2002.

[38] C. Ming-Gang, *Use of praziquantel for clinical treatment and morbidity control of schistosomiasis japonica in China: a review of 30 years experience*, Acta tropica **96** (2005), no. 2-3, 168–176.

[39] JYT Mugisha and H. Ddumba, *The dynamics of a fisheries model with feeding patterns and harvesting: Lates niloticus and Oreochromis niloticus in Lake Victoria*, Applied Mathematics and Computation **186** (2007), no. 1, 142–158.

[40] AO Nicholls, PC Viljoen, MH Knight, and AS Van Jaarsveld, *Evaluating population persistence of censused and unmanaged herbivore populations from the Kruger National Park, South Africa*, Biological Conservation **76** (1996), no. 1, 57–67.

[41] M. Njiru, E. Waithaka, M. Muchiri, M. van Knaap, and IG Cowx, *Exotic introductions to the fishery of Lake Victoria: What are the management options?*, Lakes & Reservoirs: Research and Management **10** (2005), no. 3, 147–155.

[42] SW Njoka, GRS Ochiel, and JO Manyala, *The life history and survival of Neochetina in Lake Victoria basin: Basis for biological weed control*, (2006).

[43] JV Noble, *Geographic and temporal development of plagues*, (1974).

[44] E. NSE, *First steps for managing an outbreak of acute diarrhoea*.

[45] R. Ogutu-Ohwayo, *The decline of the native fishes of lakes Victoria and Kyoga (East Africa) and the impact of introduced species, especially the Nile perch, Lates niloticus, and the Nile tilapia, Oreochromis niloticus*, Environmental Biology of Fishes **27** (1990), no. 2, 81–96.

[46] ———, *Management of the Nile perch, Lates niloticus fishery in Lake Victoria in light of the changes in its life history characteristics*, African Journal of Ecology **42** (2004), no. 4, 306–314.

[47] D.J. Salkeld, R.J. Eisen, P. Stapp, A.P. Wilder, J. Lowell, D.W. Tripp, D. Albertson, and M.F. Antolin, *The potential role of swift foxes (Vulpes velox) and their fleas in plague outbreaks in prairie dogs*, Journal of Wildlife Diseases **43** (2007), no. 3, 425.

[48] M.T.K. Tsui and W.X. Wang, *Temperature influences on the accumulation and elimination of mercury in a freshwater cladoceran, Daphnia magna*, Aquatic toxicology **70** (2004), no. 3, 245–256.

[49] ———, *Uptake and elimination routes of inorganic mercury and methylmercury in Daphnia magna*, Environ. Sci. Technol **38** (2004), no. 3, 808–816.

[50] M. Vital, H.P. Fuchslin, F. Hammes, and T. Egli, *Growth of Vibrio cholerae O1 Ogawa Eltor in freshwater*, Microbiology **153** (2007), no. 7, 1993.

[51] HB Ward, *The Bubonic Plague*, American Naturalist (1910), 439–443.

[52] J.R. Wilson, N. Holst, and M. Rees, *Determinants and patterns of population growth in water hyacinth*, Aquatic botany **81** (2005), no. 1, 51–67.

[53] J.R.U. Wilson, O. Ajuonu, T.D. Center, M.P. Hill, M.H. Julien, F.F. Katagira, P. Neuenschwander, S.W. Njoka, J. Ogwang, R.H. Reeder, et al., *The decline of water hyacinth on Lake Victoria was due to biological control by Neochetina spp.*, Aquatic botany **87** (2007), no. 1, 90–93.

[54] M.J. Wonham, M.A. Lewis, J. Renclawowicz, and P. Van den Driessche, *Transmission assumptions generate conflicting predictions in host-vector disease models: a case study in West Nile virus*, Ecology Letters **9** (2006), no. 6, 706–725.

[55] L. Wu and D.A. Culver, *Daphnia population dynamics in Western Lake Erie: regulation by food limitation and yellow perch predation*, Journal of Great Lakes Research **20** (1994), no. 3, 537–545.

[56] Y. Xie, M. Wen, D. Yu, and Y. Li, *Growth and resource allocation of water hyacinth as affected by gradually increasing nutrient concentrations*, Aquatic Botany **79** (2004), no. 3, 257–266.

Index

www.ingramcontent.com/pod-product-compliance
Lightning Source LLC
Chambersburg PA
CBHW031944170526
45157CB00002B/376